W9-BSY-881

PRAISE FOR RESTART

"Prior to the pandemic, Doreen joined me on my podcast 'Getting Curious with Jonathan Van Ness' to shed insight on how reliant on technology as a culture we are, and the subtle ways it's affecting our lives mentally and emotionally. Fast forward to now, and we are a society that IS solely reliant on tech. *Restart: Designing a Healthy Post-Pandemic Life* reflects on the power and health of our community. I highly recommend this piece not only as essential reading to get us back on track from the pandemic but as a reminder to us that it's okay to hit the restart button and to break away from distraction—especially when it comes to our own vital self care."—**Jonathan Van Ness**, host of "Getting Curious with Jonathan Van Ness" and Netflix's "Queer Eye"

"Doreen has done it again. I was not at all surprised when she started writing about helping all of us cope with, and recover from, the emotional impacts of the pandemic. Doreen is an advocate for love and the power of human connection. She shares yet another wonderful book meant to help and heal our world, one person at a time. *Restart: Designing a Healthy Post-Pandemic Life* is an incredible gift to all of us and a powerful tool in facilitating the healing process. After major trauma, life never really goes back to normal. We instead craft, and adjust to, new norms. We can decide to be part of the solution when we actively engage in the healing process, extending each other a hand of grace. We can heal the world, one relationship at a time. Through her book, Doreen teaches us that if we parent with empathy, treat each other passionately, and show kindness toward ourselves, our interpersonal spaces will become places of safety. Healing begins with love. *Restart* is a divine intervention and is the right book at the right time."—**Omar Reda**, MD, international trauma expert and author of *The Wounded Healer: The Pain and Joy of Caregiving*

"This book is a great way to reflect on what we've all been through with COVID-19 over the last year. There are so many valuable insights and actionable steps as we cautiously re-enter the world. For businesses, and people in general, it covers everything, including tips on how we can make an effort to create a safe and thoughtful space for our teams as they return to the workplace."—**Melanie Chandruang**, owner of WeConsult

"*Restart* gets right at the heart of why we've all been gutted so brutally by this pandemic, identifies clearly what we've given up in order to survive, and instructs us on how to emerge from all of it with grace. Filled with easy exercises that are rooted in the science of healing, *Restart* arms us with a bevy of useful ways to, well, restart! This is an instruction manual that teaches us how to honor our grief, heal our wounds, and build a new future that is much better than the one we all think we remember. Vital, timely, powerful, and kind, *Restart* is the gift to the world that we all need right now."—**Trystan Reese**, author of *How We Do Family*

"The past year has been brutally painful for so many of us. Our assumptive world has been upended, and some of us have lost those dearest to us. As the world opens up, we are reminded of how we became isolated, anxious, and perhaps even depressed in the first place. It's as though we have to get our sea legs all over again. Dodgen-Magee does a beautiful job at describing what we've been through with clarity and kindness, honoring all experiences and clearing a path to hope and healing. Her book provides a thoughtful examination of the trauma we've all incurred in the last year, offering ways to mindfully get back to living."—**Amy Stewart**, LCSW-S

"A book that settles your nervous system and holds space beautifully as we emerge from the pandemic, with honest, empathic, and practical guidance to support our relationship with ourselves, each other, and the world around us."—**Katie Brockhurst**, author of *Social Media for a New Age*, podcaster, and UK radio host

"Doreen Dodgen-Magee writes about grief with unique and heartfelt insight. She brings her perspective as an activist and mental health professional to write about the national grief crisis we face amidst the COVID pandemic. For years, Doreen has been a beacon of hope and resilience for so many and her new book, *Restart: Designing a Healthy*

Post-Pandemic Life will do the same. During an unprecedented time, I encourage you to pick up a copy to learn Doreen's concrete and practical advice and tips that will help survivors and allies find support and heal."—**Chris Kocher**, founder and executive director, COVID Survivors for Change; and former director, Everytown Survivor Network

"As a COVID-19 survivor/long-hauler this book provided a look into my very soul. It offers me a starting point in order to figure out how to gather all of the puzzle pieces of my scattered life as I attempt to figure out my new normal!"—**Marjorie Roberts**, PhD, life coach, COVID survivor

"In *Restart: Designing a Healthy Post-Pandemic Life*, Dr. Doreen Dodgen-Magee has gifted us with a book that offers a path forward into a post-pandemic life that feels like the 'new normal' many of us have been craving: one of embodiment, deep and real social connection, and social justice. Without minimizing COVID-related grief and despair, she offers strategies, questions for reflection, and concrete steps for adults, teens, and children to navigate anxiety and uncertainty related to emerging from the COVID-19 pandemic. I especially appreciate her guidance related to transitioning from a screen-based social life to the joyful messiness and complexity of in-person connections. Reading *Restart* made me feel hopeful and excited about the future, and that's saying a lot after such a tough year."—**Christina Malecka**, MA, LMHC, founder of Screen Time Life Line

RESTART

RESTART

Designing a Healthy
Post-Pandemic Life

doreen dodgen-magee

ROWMAN & LITTLEFIELD
Lanham • Boulder • New York • London

Published by Rowman & Littlefield
An imprint of The Rowman & Littlefield Publishing Group, Inc.
4501 Forbes Boulevard, Suite 200, Lanham, Maryland 20706
www.rowman.com

86-90 Paul Street, London EC2A 4NE, United Kingdom

British Library Cataloguing in Publication Information Available

Library of Congress Cataloging-in-Publication Data

Names: Dodgen-Magee, Doreen, author.
Title: Restart : designing a healthy post-pandemic life / Doreen Dodgen-
  Magee.
Description: Lanham : Rowman & Littlefield, [2021] | Includes
  bibliographical references and index. | Summary: "This book offers
  recommendations of how to set norms that will help readers manage
  anxiety, hesitance, and over excitement about re-entering an interactive
  world post-pandemic"—Provided by publisher.
Identifiers: LCCN 2021015941 (print) | LCCN 2021015942 (ebook) | ISBN
  9781538160275 (cloth) | ISBN 9781538160282 (epub)
Subjects: LCSH: COVID-19 Pandemic, 2020—Psychological aspects. |
  COVID-19 (Disease)—Social aspects.
Classification: LCC RA644.C67 D63 2021 (print) | LCC RA644.C67
  (ebook) | DDC 614.5/92414—dc23
LC record available at https://lccn.loc.gov/2021015941
LC ebook record available at https://lccn.loc.gov/2021015942

This book is dedicated to the families of the 2.5+ million people whose lives were taken by COVID-19 and the almost 66 million people who tested positive and survived. It is also offered, with great thanks, to everyone who did the hard work of trying to slow its spread. May we restart our communal lives honoring the memory of these precious ones by living with great intention and wise (not wild) abandon.

CONTENTS

INTRODUCTION

Early on in the pandemic, I wrote a piece for *Psychology Today* about "Zoom Fatigue." I typed it out quickly after a day that included eight hours of back-to-back telehealth sessions with clients. I did a quick read-through and pushed "publish" in the online portal without much thought. A couple of days later, it had hundreds of thousand reads and my email box was overflowing with messages of "Thanks!" "I hear you!" and "Help! What can I do to combat this?" For all the ways in which the need to socially distance had disrupted our sense of normal and calm, even in those early months, it seemed as though it had also bonded us as global people experiencing the fallout of exposure to prolonged distress with no known end in sight.

Surviving a global pandemic is no small feat. Being connected to someone who did not survive is an even bigger one. For all of us, the massive personal and global changes required to curb the spread have taken a toll. Some have felt this toll in their pocketbooks, and others have experienced the full weight of it in considering their vacuous social calendars. For many, there have been immeasurable costs to mental health. As we come to a new chapter of the pandemic, where vaccines are available and we are re-entering embodied spaces, it's impossible to imagine that we won't struggle to re-orient ourselves after experiencing such profound losses. This book is offered as a guide as we engage this necessary struggle.

In many ways, our pre-pandemic, symbiotic relationship with our devices brought us into the COVID-19 landscape bereft of the very skills that would have helped us endure it. We entered quarantine unable to focus, delay gratification, or be bored, hyper-stimulated, feeling constantly distracted, comparing ourselves to the social media presence of others, and, basically, out of touch with our deepest sense of self. This left us largely unable to handle the isolation, boredom, and uncertainty that quickly became a part of daily life. At the same time, we had the sense that technology could save us . . . and, in many ways, it did. Quickly and fully migrating our embodied lives over to digital spaces allowed us the incredible gift of keeping going. It also, quickly, exhausted and overwhelmed us.

Given that excessive technology use can erode our abilities to think critically and self-soothe, it's likely that most of us greeted lockdown orders with deficits in both. As the pandemic raged on with no clear end date, the onslaught of news about political, racial, and global upheaval overtook our lives and attention. Overwhelmed, overstimulated, out of control, and bereft of access to our normal "coping" strategies and routines, our flagging abilities to focus, evaluate, and self-soothe were difficult to access. Which brings us to this moment, where we find ourselves exhausted, emotionally and physically dysregulated, and correspondingly thrilled and terrified of how we will ever find a new sense of "normal" amid the wreckage of all we've lost in this time. COVID has touched every part of our world and our lives. At the time of press, almost 2,500,000 lives have been taken by COVID-19 around the globe and countless others have had the virus. It's nearly impossible to find someone whose life has not been touched or who is not facing unprecedented losses as a result of the pandemic. We've rarely grieved globally as we have this year.

Complicated grief is defined by the World Health Organization as "a persistent and pervasive grief response characterized by longing for the deceased or persistent preoccupation with the deceased accompanied by intense emotional pain (e.g., sadness, guilt, anger, denial, blame, difficulty accepting the death, feeling one has lost a part of one's self, an inability to experience positive mood, emotional numbness, difficulty in engaging with social or other activities)."[1] In this year-plus of losing nearly every experience of normalcy, it's fair to say that many of us are locked in a complicated grief cycle, caught up in ruminating over losses, worrying about the consequences of them, and desirous of excessively avoiding reminders of the carnage of the year.[2] We'd love to act as though nothing happened and leap right back into the wide world, while at the same time sensing that we've been forever changed.

Not only are we grieving but we are also living with personal and cultural trauma that won't just live in our pasts. *Trauma* is defined by the American Psychological Association as "an emotional response to a terrible event like an accident, rape or natural disaster. Immediately after the event, shock and denial are typical. Longer term reactions include unpredictable emotions, flashbacks, strained relationships and even physical symptoms like headaches or nausea." Given that the experience of trauma stays with us and is activated in physical and emotional ways long after the original situation is over, it's fair to say that we will all be feeling the effects of prolonged exposure to cultural and individual trauma in the coming months and years. When human error is involved in the traumatic situation, such as with the withholding of information or failure to provide instruction on how to stop the spread of the virus, the complications of the grieving and traumatic response increase.

We will all need extra measures of grace and kindness toward ourselves and those we interact with as we work through both our communal and individual griefs and traumas. Each of us has been impacted in unique ways as well as in ways that we share. Every person's response is likely to look a bit different from the next person's. What is likely to be common, however, is the felt fragility of a new chapter, the exciting and terrifying opportunity to connect with those people and experiences we love, and a desire to never experience quarantine in this way again.

The following chapters are intended to give brief but thorough analysis of where we have come from and ideas about how we might move forward with health and thriving in mind. "We'll all get through this together" became a familiar rallying cry over the last many months. Let's keep it alive now. With some forethought, planning, and intentional work, we can move forward together to support each other's mental health, create rewarding embodied realities, and nurture healthy new relationships with each other and our devices.

I

THE LAY OF THE LAND
AS WE RESTART

1

WHERE WE'VE BEEN, WHERE WE ARE, AND WHERE WE'RE GOING

Shortly before the world shut down, Shay moved across the country with her two children. Recently divorced, she and her former partner had agreed that being near Shay's extended family would be best for Shay and the kids. Upon arriving in Michigan, Shay found a part-time job as a massage therapist at a medical clinic and the kids began attending a new school. Two weeks later, Shay was furloughed due to stay-at-home orders, and schools were closed for in-person instruction. Although the children encountered their classmates via distance learning, they weren't included in out-of-school, online socializing because they hadn't fully connected with classmates in person before lockdown. Connection with Shay's elderly parents was halted because of the high risk of catching the coronavirus, and navigating the need to establish new relationships with pediatricians and other medical professionals in a new town became impossible.

Quarantine-related isolation became very real very quickly—and Shay and the children became alternatingly clingy and frustrated with one another. Due to travel bans, summer came and went without the children getting to see their other parent. With no one and nothing to offer a break from the monotony, the acute awareness of being isolated in an unfamiliar community became stark. In turn, all three felt passively sluggish and actively disgruntled as they entered a new school year.

Now a year into the pandemic, Shay and her kids still feel isolated and are exhausted, overly reactive, and depressed—with no sense of home or community. Shay is still unable to see massage clients, feels the financial impact of her furlough, and worries about how she will continue to support her family. The clinic where Shay worked is unsure if they'll even retain her services "post vaccine." And the children are still hesitant about returning to school in person because they've only had contact with their classmates online and don't feel as though they fit in. While the world around them rejoices in the thought of returning to "normal," Shay and her children feel acutely aware there is no normal to return to.

Shay's pandemic challenges may seem more extraordinary than yours or mine—and her story isn't even the most complicated one to unfold during these trying times! But comparison is not the point. Sure, it's wonderful to be able to say, "I have it easy compared to X person," but it's also crucial to own the ways this time has made its personal mark. Every one of us has experienced losses and traumas of all kinds and sizes. We've missed proms and graduations, postponed weddings and momentous trips, and said heart-wrenching goodbyes to loved ones. We've fallen behind in schooling, long-planned adventures have gone unexperienced, and much of what we've lost can never be regained. The traumatic effects of these losses are encoded in our bodies, minds, and hearts. To resolve the trauma, we need to be consciously aware of our personal losses. Once we are aware, we can be intentional about the way we tend to ourselves and our emotions around what has been and what is to come. Rather than denying the traumatic losses, we can address them in ways that might help us to heal.

When I speak on the topic of trauma as a psychologist, I often use this analogy: Life before trauma is a lot like walking through a field on a sunny day. Enthralled by the beauty around us, we notice the clouds and hear the birds. The feel of the grass grazing our ankles is cool, and the wind feels refreshing. As we meander, we notice a small coiled object to our side but don't think much about it, as all is safe and well. When the trauma hits, we suddenly feel shooting pain emanating from our ankle and adrenaline floods our system. Emotionally, we panic and default to a fight, flight, freeze, or faint response. Looking down, we see we've been bitten and that the small coiled object is actually a snake. In that traumatic moment, our body and our brain encode the experience in unique ways that will be triggered when we are in a similar situation. When we walk around in the world and notice any coiled object, our body responds as though it's a snake even if it's a coiled garden hose.[1] When we've experienced trauma, we don't have the cushion of a pause to determine how we respond; we just respond.

Very likely we all have some trauma stored in our brains and bodies as we depart quarantine. That trauma may feel small and minor, or it may feel large and primary. The trauma also may flux in and out of our awareness. Some of us may feel it around all we've missed out on during this time of distancing. Others may feel it when endeavoring to leave home, fearful we could still transmit or contract the virus. Every reaction is normal, and it would be impossible to list all the kinds of distressing and traumatic reactions we may encounter in the coming months.

The following sections in this chapter highlight changes in various areas of life and experience over the course of the COVID-19 pandemic. Although all of us may not have experienced each of these, very likely someone in our community has been impacted by the particular shift that occurred as we all endeavored to stay safe from the virus. In order for us to "restart" post-pandemic and re-enter the world in healthy ways, it's important we reflect upon our own pandemic experiences as well as those of others.

TOILET PAPER HOARDING

One of the first national responses to stay-at-home orders in the United States was the disappearance of toilet paper at nearly every store. Granted, other items quickly became hot commodities. Marinara sauce and bread were hard to find; for a while, procuring eggs was a challenge. Toilet paper, however, was simply gone. What an important barometer of where we were when the pandemic hit!

Embroiled in cultural and political binaries, facing the most important election in modern history, we Americans were told to stay at home at a time of great unrest. Having faced four years of divisive leadership, half the population felt traumatized and gaslit while the other half felt emboldened and validated. No one had much margin or cushion left emotionally.

Into this space came massive fear of the virus—followed by a fast reveal of the systemic lack of preparation for a pandemic, overwhelming numbers of COVID-19 cases, overcrowding of hospitals, and a devastating number of deaths. When we are out of control and feel as though we are falling, we will reflexively grab at whatever we can find to give us a sense of sure footing. Never mind if what we grab is only a door handle attached to nothing.

In the early days and months of the pandemic, toilet paper was the thing to grab. We heard it was selling out, so we bought more than we needed, and we did this every time it appeared on the shelf. Reflexively, we thought only of ourselves and our comfort and it showed in our hoarding of this staple. Eventually, the manufacturers caught up with increasing demand and we all stopped panicking.

Think about our pandemic toilet paper obsession as a window into our cultural reality. We were a people divided. We were either terrified or dismissive, obsessed with the numbers of ill and dying or with uncovering the whole thing as a hoax. In our stressed-out, divided state, we found ourselves out of control. It isn't toilet paper that's the issue here; it's that we wanted to control something, anything—and hoarding something rather than buying only what we needed, as we needed it, seemed to feed at least a small part of that need to be in control. Without reflection and exploration, we risk perpetuating this kind of behavior in large and small ways. Wouldn't it be better to think through why we behaved in such a way and learn from the experience?

RESTART WITH SCARCITY MINDSET EXPLORATION

- **Identify times you've missed out because you weren't hyper-vigilant or fast/dogged enough in your pursuit.** When in your life have you experienced competition for a resource or object?
- **Explore possible scarcity mindsets.** When you're thinking, "There isn't enough for everyone," you are experiencing what's called "scarcity mindset." In times of prevailing group think geared toward a scarcity mindset, how do you feel? Scared? Angry? Competitive? Depressed? Hopeless?
- **Name the ways in which the pandemic triggered scarcity fears.** In what ways and to what extent have scarcity feelings been activated for you during this time of stay-at-home mandates and social distancing?
- **Make and practice a mantra.** What are some mantras that could replace scarcity ones as you "re-enter" the world? Examples include, "I will have what I need when I need it," "I'm part of a community that will share," "I can be creative about getting my needs met," and "I can grow my capacity for empathy by considering the needs of others alongside of my own."

THE WIDENED OPPORTUNITY GAP

Moving our lives to online spaces during the pandemic has been easy for some and impossible for others, often because of access to resources. Those with greater resources—money, time, connections, education, and more—are always set up for a greater chance of success in any situation. Economically advantaged people usually have personal connections who can support them in their adaptation to challenging times. Plus, they can pay for any needed support. They have bigger living spaces, where working from home is more than just possible; it's easy. They likely have access to better internet and newer, faster electronic devices. Children in these families may benefit by having the opportunity for more input from their parents, who can afford to step away from work or to hire other support people such as caregivers, pod leaders, or tutors. These kids go to upper-tier schools that can similarly afford to pay for exceptional staff, high-quality mobile learning opportunities, and robust social–emotional support resources. On the other hand, students from lower-income homes often have fewer resources to help them cope with their relational needs. They also usually have lower-quality technology and internet, making online learning nearly impossible. We have created a system that stacks the deck against them.

Privilege is defined as "the unearned social, political, economic, and psychological benefits of membership in a group that has institutional and structural power."[2] Society is, in many ways, built upon systemic patterns that privilege some people and oppress others. When those in privileged subgroups don't own their access to substantial advantages, they can perpetuate unjust resource distribution. For an eye-opening experience, take the privilege assessment found at https://www.buzzfeed.com/regajha/how-privileged-are-you. As we re-enter communal life, it's important that we do so with our eyes wide open to the ways this year has impacted us disproportionately.

Specifically, much of the success, or lack thereof, of people's vocational, educational, and relational pursuits during quarantine has relied upon robust and reliable online connection via devices that can be counted on to work without glitches. For this reason, those without such resources have suffered greatly throughout the pandemic. Students without access to reliable technology and employees who could not adapt and work from home are facing particularly difficult inequities in the way of opportunities. With many restaurants, arts organizations, nonprofits, travel-related businesses, and more facing permanent closure, hundreds of thousands of employees may not even have a job to which to return! Similarly, students who have fallen behind due to lack of resources or support for learning face mounting difficulties in rejoining their pre-pandemic classes.

Although some lost ground might be regained when in-person learning and working return, we must acknowledge that those with fewer resources have weathered this time with greater challenges and less relational connection and support. Yes, the gap between the haves and the have-nots existed long before the pandemic, but that gap is now a vast chasm. Without reliable technology and supportive resources, too many people have been completely left behind during the pandemic. This same demographic also has experienced the most losses to the virus itself. We must attend to this reality.

RESTART WITH GREATER EQUITY

- **Wake up to personal privilege and build empathy**. Once you have done at least one thing to understand your own positions of ease, consider how you might take one action to empower (not "save") people who have greater challenges accessing resources.
- **Identify those in your community who could benefit from contributions or volunteers**. Offer with humility and learn about ways of giving to these communities that don't center yourself.
- **Work through unconscious bias**. What kinds of biases do you have that may inhibit your empathic care for vulnerable populations? Harvard has an incredible implicit bias assessment that can help you identify blind spots. You can find and take it here: https://implicit. harvard.edu/implicit/takeatest.html.
- **Get help. Listen and learn**. Who do you know or what do you have access to that might help you work through these biases in order to respect the reality of the needs of these communities?

VULNERABLE POPULATIONS
MADE MORE VULNERABLE

Black, brown, and indigenous people have disproportionately contracted the virus and died from it.[3] This horrific reality will have a lasting psychological impact on everyone in those communities and should wake all of us up to the injustice and unfair distribution of resources in our communities. In addition, incarcerated individuals and their families have been profoundly scarred by the virus, as have seniors living in nursing homes.[4] The systemic injustices these vulnerable populations experienced prior to the surging virus, piled on with fewer protections for them during this time, likely means the way they re-enter embodied life may reflect the trauma of inherited inequity.

People who live in unsafe settings have also faced unique challenges during the pandemic. Specifically, women living with abusive partners have had few options for exiting these relationships and have, likely, faced intensified abusiveness given the pressures inherent in this time.[5] Likewise, houseless individuals have suffered severely; many of the resources that enable them to survive—from food pantries to shelters—could not safely operate during sheltering-in-place orders and literally have left people out in the cold.

Children, who are often vulnerable simply because of their age, have faced challenges as well. Because a child's ability to manage distress is largely dependent upon the unique blend of temperament, caregiver capability, and available resources, children will emerge from this time with varying levels of vulnerability. Children with less guidance from wise and healthy adults—including teachers, arts instructors, sports coaches, and the like—will be at risk of re-entering the world with academic, extracurricular, and emotional deficits. Those who previously relied on nutritional assistance at school may be paying additional physical costs, as might kids who haven't seen a dentist or a pediatrician for their annual checkups. Children who have been quarantined without access to reliable internet or digital devices may re-enter with fewer social connections and weaker academic skills. Finally, children who have been living with unsafe adults may be facing harsher physical and emotional treatment, and possibly abuse, during this time.

As we move forward individually and collectively into these next phases of the pandemic, we all need to pay special attention to these vulnerable populations and do what we can to increase their chances of thriving in a post-pandemic world.

RESTARTING FOR VULNERABLE COMMUNITIES

- **Take stock of how this time has impacted you as well as those in your community.** Identify your greatest needs in this time as well as those of your community. Think of what you may need to ask for and/or what you might offer. Challenge any "bootstrap" thinking. While you may have had to pull yourself up by your bootstraps in this time, remember that there are others who had no boots. Lead with empathy rather than comparison and judgment.
- **Take stock of your support systems.** Going through the contacts in your phone, identify anyone who fits into any of the groups described above and consider ways you might reach out to them to offer empathy and/or support.
- **Get generous.** If you aren't a member of one of these COVID-19–impacted communities or know anyone who fits into the vulnerable categories, consider making a monetary donation to a nonprofit organization serving Black, brown, or indigenous people, a women's shelter, a food pantry, or a Head Start or early childhood program in your area. The need is monumental.

THE IMPACT OF CONNECTION THROUGH VIDEO

Before COVID-19, the thought of migrating nearly all of our interpersonal interactions to video platforms was a sci-fi plotline. Very quickly, however, it became our reality and, with it, all kinds of new social dynamics that offered advantage to the relationally comfortable and housing secure and disadvantaged everyone else. This happened in a multitude of ways.

One of the advantages of shared work spaces is that they provide a neutral environment for people to gather in. People living in less-than-ideal surroundings can work and learn safely when they have a job or classroom to go to. This has a real and profound impact on a person's ability to be present for their work and/or learning. When these same individuals are required to have their homes on display, things can feel extremely different. Not only might they have a less-than-ideal work or school environment (due to surroundings or the presence of others when they are working and learning at home), but they also may feel judged or evaluated based upon the surroundings that show up in the video format.

Video platforms, in and of themselves, also offer glitches that alter the ways in which we communicate. Those of us hesitant to speak out in a group have likely spoken much less in these digital spaces, where it can feel difficult to understand and navigate the rules of engagement. Because most video technologies can broadcast only one voice at a time, those who talk loudest and most assertively have been able to easily "grab the mic," whereas those who are more hesitant often lose it. We've also had to try to fill in gaps to our understanding since in-person energetic exchanges help us decipher how a particular interaction is going. These dynamics, and more, have made a profound stamp on how we behave individually and in groups. This is unlikely to go away instantly.

Another phenomenon of video interaction is that of having our own faces in front of us for a majority of the time we are communicating with others. Pre-pandemic, if someone would have suggested we carry a mirror in which to watch ourselves whenever communicating with others, we would have named this a horrible idea. For most of us, reflecting on our own image leaves us less likely to fully tend to the person with whom we are speaking and more likely to be highly self-evaluative. This is the biggest reason video calls are so exhausting! Yes, they rob us of information forms such as nonverbal cues and they remove the interpersonal "energy" that is more easily transmitted in embodied get-togethers, but this reality of constantly evaluating how we appear has taken a real toll. Although for some it may be a nonissue, for a majority of us, seeing and evaluating ourselves all the time has been exhausting and has cost us in the way of self-confidence and spontaneity.

RESTART WITH VIDEO CALLS

- **Identify your feelings**. Consider the ways you feel during video calls. Self-conscious? Empowered? Relieved not to be in-person? Sad not to be in-person? If you have more than one video call in a row, do your feelings intensify?
- **Realize changes and make the unconscious conscious**. How have your feelings about your appearance changed over the course of the pandemic? Do you feel comfortable or uncomfortable when you see yourself mirrored as you talk online? Approximately what percentage of your attention goes toward scanning the image of yourself and adjusting your communication as a result? Begin addressing your self-consciousness and find ways of keeping it in check.
- **Do some video call self-care**. Using the answers to the questions above, begin to make a self-care plan that includes being compassionate with yourself. Items to include might be:
  - Spruce up a space specifically for video chats and have that be your go-to location.
  - Before heading on camera, take a minute to look in the mirror, make any adjustments, and affirm yourself and your appearance.
  - Place a sticky note near your screen that affirms what you need. That note might read, "I am capable," "I know what I'm doing," "It's just me that's not used to seeing my face," or whatever words empower you.
  - Hide your view of yourself. If the platform doesn't allow you to "hide" yourself, block your device camera with a sticky note or a piece of tape.
- **Practice in-person interaction when possible**. Donning your mask, go into the coffee shop and order your coffee face-to-face or buy your groceries in person, making small talk with the cashier. Meet a friend for a distanced or safe conversation rather than doing it via video or text. Practicing soft social interactions will help us as we tackle the bigger ones required for a full restart.

THE PRICE OF OVER-RELIANCE ON SCREENS
(WE ARE DISTRACTED!)

In many ways, social media has saved us, allowing us to stay in touch with other people. In other ways, however, it has added to our screen fatigue, our fear of missing out, our physical lethargy, and, at times, our anxiety about the world. In fact, research shows that excessive social media use is not only related to anxiety and a fear of missing out, but can actually cause those psychological realities.[6]

The simple acts of leaving our home for work or school, or our desk between meetings, created pre-pandemic rituals of sorts; we were required to move our bodies and shift our attention to things other than our screens and social networks. During pandemic lockdown, however, as our embodied and digital lives overlapped even more profoundly, we too often simply stayed at our screen during breaks. We lost starting and stopping behaviors and, instead, simply toggled out of our work or school screen to social media or gaming on the same screen.

Without places to go, we all began bonding over experiences we could share in digital spaces. We all began baking. Banana bread first, then sourdough. We all binged *Tiger King* and lived life playing *Among Us*. We got wrapped up in the 2020 election and had alerts set for everything. We were distracted and we broadcasted and connected over it all in our social media spaces.

This constant connection to both curated images of others and ever-evolving information existed before the COVID-19 pandemic and has only intensified throughout quarantine. This means we are likely re-entering embodied living with less ability to relate authentically to others and to focus and regulate our emotions than we had before quarantining. We've been able to toggle between things, and we've developed a proclivity for it—but doing so impacts the way our brain can help us get settled and focus on one thing at a time. We've also overstimulated ourselves, when what we really thought we were doing was soothing ourselves. Endless hours of checking social media on every break and bingeing the latest show in every off work or school hour has not brought us to a state of emotional regulation like we think. It's simply distracted us from our emotional dysregulation—which is costly.

RESTART WITH ONLINE BEHAVIOR

- **Assess your dependence on your devices.** Look back at your online viewing, web browsing, and device-use history, or simply count on your memory to consider how much and with what tech you've engaged most heavily during quarantine. Are you comfortable with your use?
- **Notice what your device use tells you about your emotional state.** Given that the content to which we're drawn tells us something about how we're feeling, what does your use tell you about what you needed or leaned toward in this stressful time?
- **Evaluate how your tech use helps or hurts your emotional well-being.** In hindsight, did the content you consumed help you to cope with the challenges of the COVID-19 pandemic, or did it increase feelings of being overwhelmed? Were you able to find calm and centeredness during the pandemic, or did you substitute distraction for taking care of your emotional needs?

Turning around our cognitive distractibility and emotional dysregulation is likely going to take some dedicated time and concerted effort. It is imperative, however, that we do the hard work. (Re)learning to actually soothe ourselves and our central nervous system—and to be able to maintain focus on a particular situation—will reap big payoffs as we restart.

RUNNING A MARATHON WITH A CONSTANTLY MOVING FINISH LINE

Somewhere around the sixth month of quarantine, I wrote an article on what researchers call the "Third Quarter Phenomenon (TQP)."[7] I was convinced, at that point, that the psychological symptoms I was noticing in myself and others were related to where we were in the timeline of the pandemic, and I was certain we were nearing our final span of quarantine. Little did any of us know how wrong I was.

TQP involves a set of psychological symptoms commonly experienced by people whose work requires them to live in isolation for a preset period of time. Astronauts, submarine workers, researchers in sub-Arctic climates, as well as those in solitary confinement all fit into this category. For each of these groups, TQP shows up when people are just about a quarter of the way through their time in isolation. Agitation, irritability, depression or fluctuations in mood, and decreased morale comprise the set of psychological difficulties that define TQP. Research in this area shows that people cope with extreme isolation without experiencing symptoms in the early stages but don't do so well as time goes on. With an end date in sight, they begin to long more actively for relief and develop symptoms when it doesn't come immediately.

These same psychological symptoms have plagued most of the global population for a large part of the quarantine. It makes sense that having our freedoms limited, while also needing to maintain specific safety guidelines, all while losing access to most forms of distraction and fun certainly would take a toll on morale as well as on our personal and communal mental health. When we hunkered down to shelter in place, we had no idea how much we'd need to pace ourselves to get through this time unscathed.

TQP research illuminates how people who are somewhat prepared for the rigors of being cut off from normalcy often experience distress. The pandemic offered this same reality completely outside of the research setting and with no preparation. For all intents and purposes, the pandemic has been a marathon for all of humanity to run. Things may have turned out differently for us if we had been prepared for what was to come. Many of us started out in full sprint mode, with little to no training or forethought on how to sustain ourselves for the long haul. For those who began the COVID marathon with preexisting stressors, mental health needs, or an overarching tendency toward hopelessness, things felt impossible right out of the marathon gate. For those who entered quarantine energetic and positive in their efforts to create new routines and practices, maintaining this posture became difficult as the time went on with no known finish line.

The impact of this finish-line-is-always-moving reality will be felt for a very long time. For some this will be evident in a nagging fear of the "other shoe dropping." For others it will be a hesitance to trust the reliability of the vaccine and a tendency to continue to isolate even when it's not required or to experience heightened anxiety in response

---

RESTART WITH OUR OWN DISTRESS AND/OR TRAUMA

- **Assess pre-pandemic experiences with trauma and distress.** How did you handle prolonged distress before COVID-19? What trauma did you experience before the pandemic? This might include significant loss including, but not limited to, death and divorce; experiences when you were harmed; serious physical illness; or other hardships. If you work in a healing profession; medical or environmental crisis-response sector; or are a volunteer, organizer, or activist, secondary trauma resulting from exposure to the trauma of others also matters.
- **Identify tools that helped.** What measures or steps have you taken to explore and work through the impact of your trauma?
- **Clearly identify personal distress and/or trauma during quarantine.** In as specific and concrete terms as you can, list how quarantine life was distressing or traumatic for you. Did you feel out of control? Fear the unknown? Lack access to regular routines and activities? Lack physical contact with others?
- **Name specific remaining distress and resources for addressing it.** What emotional scars remain as a result of exposure to these experiences for a prolonged and unknown period of time? Once identified, who are the friends, community members, agencies, or professionals who could help you deal with your trauma?
- **Commit to actively address self-care.** Make an appointment on your calendar to tend to one or more of these unresolved traumas in your life. Read a book, see a therapist, talk with a trusted friend, or journal. Keep that appointment!

---

to being out in public. For some it may show up as a complete disregard for continued safety measures.

Despite the unique ways these impacts will manifest in people, we are all experiencing varying levels of distress and/or trauma as we emerge from the pandemic. Just as we would likely experience physical injuries and psychological distress if we undertook a marathon with no training, no preset course, and a moving finish line, so, too, we are experiencing the reality of a prolonged period of distress filled with difficult and threatening unknowns.

THE MAJOR HIT ON EMBODIMENT

Years ago, I took my nephew bowling. He had never been to a bowling alley, but he had, just the prior Christmas, been gifted a Wii and loved its bowling game. I talked up the in-person bowling experience heavily. I was excited to introduce him to the sensory experience of rolling balls, crashing pins, jukebox music, and musty rental shoes. As we walked through the doors, Ethan stopped dead in his tracks. With his eyes as big as plates, he exclaimed, "Auntie! They made it look just like Wii bowling!"

One of the effects of our excessive reliance on technology is a new kind of relationship with our bodies and the physical spaces they invade. Rather than Ethan realizing that Wii was approximating embodied life, he thought the opposite was true. This makes sense, because the innovative, over-the-top way we can have experiences in digital spaces often makes the embodied ones feel like weak substitutes.

With so many of our experiences migrating to digital spaces during quarantine, we have overstimulated our visual and auditory senses while largely ignoring the other senses. When COVID-19 safety guidelines

---

RESTART WITH OUR PHYSICAL ACTIVITIES

- **Identify the physical experiences we've neglected or not been able to have during quarantine.** Make a list of the physical activities you had to set aside when the pandemic hit.
- **Rank your list.** Take at least five full minutes to rank your list from "Things Most Missed" to "Things Missed Least."
- **Make a realistic plan and stick to it.** Begin to make a plan about how to do those things you missed most as soon as your county and state officials reopen activities. Pre-planning will allow you to make decisions about how to use your time and energy with intention. If going to the gym is first on your list but you're being flooded with requests for social gatherings, being able to look at your list will help you to say no to the wonderful, but not aligned, option for yourself. Refer to this list often as you establish new routines with your body.
- **Try at least one new embodied activity and sensory "stimulant."** Identify which of your senses could use some attention (new tastes, new smells, new textures to touch) or a new physical activity that you've never tried and give it a whirl with graciousness and an open mind.

---

highlighted that being in social spaces with our physical bodies created undue risk, we found ourselves sitting in front of screens more and moving less. We found ourselves texting more and talking less. Even the mundane opportunities for soft forms of social connection such as going to the grocery store and gym were taboo, so we were left with ourselves and our screens.

Given our discomfort with boredom and our flagging ability to be with ourselves without feeling edgy, this led us to loop back to our screens for distraction and stimulation in dependent and repetitive ways. As a result, we've likely lived less in our bodies than we did pre-pandemic.

THE LOSS OF RITUALS AND ROUTINES

Rituals are important to our mental and physical health. Brushing our teeth as a part of a morning or nighttime routine, playing a weekly pick-up basketball game, and celebrating birthdays and anniversaries are all examples of rituals that help us mark the movement of time and meaningful moments in life. During the pandemic, we all created ritual behaviors related to hand-washing and mask-wearing, and those working and living in frontline embodied places such as hospitals and grocery stores have had to adjust their daily rituals to ensure social distancing, masking, and more. On the whole, though, most of us have lost significant large and small markers of time and meaning.

When I was seeing clients in person before the pandemic, I routinely stood up to accompany each person out of my office and to head to the waiting room to greet my next client. It's easy now, with my therapy sessions online via a telehealth platform, to wrap up a session, type a few notes, then migrate to email and work there until my next session begins. In losing the transitional movement and space, I have also lost the perspective shift that came with a more marked beginning and end of each appointment.

Many of us have experienced similar losses of ritual over this unique time. We aren't heading outdoors, or at least down the hall, for leisurely lunch breaks. We're not grabbing a cup of coffee in a break room or gathering around the water dispenser, catching up with colleagues. Students aren't walking from class to class, getting lunches out of lockers, or heading to recess or intramurals. All these changes have impacted the flow of life.

Even the ritual of getting dressed has changed all around the world. Loungewear has become the norm, and it's common joke fodder to refer to being dressed for meetings up top while wearing sweats or bike shorts below the camera. While dressing in comfortable clothing is, in no way, a harmful new ritual, it's simply important to note the changes that will accompany an opening up of embodied and social living. Many female-identified people tell us they never plan to wear underwire bras again, and all genders might feel better wearing elastic-waist pants a majority of the time. Will, however, our places of employment and learning adjust to these more casual clothing regimens, or will we all be forced to return to business wear and uniforms, to previous rituals around clothing?

We also face the need of re-thinking all manner of other rituals. A nonprofit for which I volunteer has historically gathered thousands of

---

RESTART WITH RITUALS AND ROUTINES

- **Identify the large and small rituals that mark time and space in your life.** Take time to identify the large and small rituals and routines you left behind during the pandemic. This could be as small as "going outside to commute to work each day," "getting out of my sweats," to as large as "completely stopped traveling, which was a huge part of my work."
- **Understand the meaning behind each ritual.** Looking back over each lost ritual, consider what that ritual did for you when you used to practice it. Did it help or hinder your health and happiness?
- **Clearly identify new rituals birthed during COVID-19.** What new rituals have you begun during the time of quarantine? Sleep in until moments before needing to start work or school? Online grocery shop? Binge-watch TV series while you eat lunch or dinner?
- **Assess how the new rituals help or hurt.** Reflecting on your new rituals, how has each helped or hindered your health and happiness?
- **Identify re-entry rituals that might help mark the transition to a new time and space.** What new re-entry rituals might help you to deal with all the changes that are to come? Get as specific as possible, write them down, and post them where they're easily visible as you work to enact them. Share them with others who can help you get them in place.

volunteers every year for an important conference and training opportunity. Having moved the event to a digital space during COVID-19 quarantine, will the organization return to the expensive reality of conference center rentals and airfare? With corporations having learned that online training works, will the ritual of in-person corporate onboarding in an actual space with embodied people go the way of the corded phone and be obsolete? Schools, businesses, and even leisure-based enterprises such as yoga and dance studios that invested in digital infrastructure to survive the pandemic will have to make the difficult decision whether to return to in-person classrooms and work spaces, to stay online, or to offer a hybrid model. If many choose to remain in digital spaces, the rituals by which we work and learn and play will be forever changed. Thoughtful consideration of the cost of this would benefit us all.

2

THE COMPLEXITIES OF
RE-ENTERING EMBODIED LIVING

All of us are embarking on the difficult task of decision-making and tending to our physical and mental health as vaccines are making their way into arms and we re-enter embodied living. We arrive at this transition exhausted, anxious, and edgy. Just as new parents undertake the massive job of learning to care for a child while experiencing the least amount of sleep and self-care they've ever faced, we're stepping into the unknown with high expectations at a time of great depletion, anxiety, and exhaustion. How will we move forward given our weary state?

The reality is that each of us will face a unique set of psychological challenges as the need for strict quarantine recedes. Our responses to those challenges will depend on our individual temperaments, personalities, and very sense of self. Some of us will face decidedly greater levels of complexity in the transition. Those of us living with others, for example, will have to confront the unique challenges and celebrations involved in our own re-entry while simultaneously tending to the needs and preferences of those we care for or with whom we live. The same complexities will exist in the larger communities in which we reside. This will certainly result in all kinds of complicated feelings and experiences.

Although the majority of the complexities we'll face in maintaining our health while leaving home to enter the wide world will be specific to each person, there are some psychological realities that will play out universally. These are laid out in this chapter, along with some exercises to help mitigate the difficulties that come with them.

DEALING WITH REACTIVITY AND INTERNAL BIAS

We all have biases, both conscious and unconscious, that lead us to think and behave in certain ways. Yet, although our biases may be different, we all share the human tendency to compare our ideas and behaviors with those of others and judge them as "sensible and right" or "flawed and wrong." The social and political tensions at work throughout the pandemic have commingled with this tendency, creating a particularly reactive environment within which we will be making innumerable decisions and taking action as we re-enter the world. In essence, there have been as many unique reactions to the pandemic as there are people in this world, and this has resulted in a lot of internal and communal discontent and conflict.

Just as we faced ever-changing news and information in the early days of the pandemic, the days of re-entry are going to be a bit topsy-turvy. Each of us is going to be required to make tons of decisions based on what we know at the time and what sources we trust to give us information. Will we choose to eat inside restaurant dining rooms and go back to the gym? At what point will we return to in-person church services or parties inside? Will we ask people to mask when sharing space with us indoors if we don't know their vaccination status? There will be so many choices to be made.

To navigate these decisions healthfully, we will need to: (1) have reliable data, (2) evaluate our needs against that data, and (3) take action in line with the resulting analysis. Giving ourselves sufficient time and offering adequate energy to do our research is going to be key to ensuring we don't burn out or make unwise, reactive decisions. The bottom line is that we need to vet our sources for the highest quality, scientifically sound data and use that as the basis of our decisions on how to best act. Sometimes we'll get decisions right and sometimes we won't. Either way, we can't expect ourselves to simply jump back into "normal" pre-pandemic life. As an added note, those of us who find ourselves ruminating for long periods of time before taking action will need to be aware of this tendency and find ways of supporting quicker decision-making so we don't add extra stress to the process by fostering self-doubt or feeding decision-making paralysis.

In addition to mindfully setting aside time and energy for making fact-based, intentional decisions about re-entering embodied life, we need to create a framework for dealing with the big feelings that result when we hear about the decisions and see the actions of others during this time. Creating this framework involves remembering and honoring that how any one person is emerging from quarantine does not define or predict how anyone else will do so. There is an old saying, "Be kind to everyone, for theirs is a difficult journey." Living with such a frame of reference can be challenging, but doing so may ensure the most community-friendly re-entry possible.

Throughout the pandemic, we have seen conflict separate friends, families, and communities—with many judging others for how they complied with or disregarded safety protocols. When we bump up against other people's choices as the pandemic winds down, we are likely to have strong reactions to what we see in their behaviors and attitudes. Too much energy expended on the behaviors of others robs us of time and energy we could more effectively spend on ourselves. When we find ourselves being reactive about others' choices, we might benefit from asking ourselves, "What is my reaction to this person's decision-making telling me about myself?" "Would this person be open to sharing with me the reasoning behind the decision, and, if so, would I be open to hearing it?" "Is there reliable and trustworthy data I might be able to provide if this person is open to it?" "Can I afford the cost of conflict if I offer it, and they are not open to hearing it?" The answers to these questions will help us determine if and how we might respond to our strong feelings about the decisions of others.

Such challenging assessments are made more complex by how personal pandemic-related choices *impact* others—especially if those choices include a disregard for established safety protocols. If our willingness to go mask-less impacted only ourselves, it would be one thing. The reality is, however, that such a choice puts those we encounter at risk and communicates that we don't really care what happens to others. Whenever we reach out to shake someone's hand, offer a hug, break through the six-foot bubble, or forego mask-wearing, we are infringing on others' rights. Although personal rights and agency are important, in this time of re-establishing communal connection, empathy and being a respectful neighbor are more important than ever! Understanding and caring about the way our personal choices and actions impact others are crucial to an evolved society and self.

BUILDING EMPATHY:
THE DANGER OF A SINGLE STORY

To be sure, this pandemic has taken a profound—and unique—toll on everyone. While some have been hit extremely hard and lost multiple family members to the virus, others have been relatively undertouched by the ravages of the COVID-19 virus. While some contracted the virus and are living with debilitating, long-term health complications, others who tested positive had only mild symptoms and have moved on with their lives. While some communities are emerging from the worst of the pandemic under the weight of massive loss, other communities have almost no experience with the consequences of COVID-19.

In her 2009 TED Talk, novelist Chimamanda Ngozi Adichie explains how hearing only one story about a person or culture can lead to forming seriously erroneous misunderstandings about that person or group. In Adichie's case, a single story about Africans led her college roommate to make many inaccurate assumptions about her.[1] The same reality exists for many who have only been exposed to a "single story" about the COVID-19 pandemic. Some may have no connection to the stories of those who lost loved ones or personal health to the virus. For those of us who fall into this category, we need to understand we are "re-entering" from a place of great privilege. We must seek stories that include the specific kinds of profound heartache and emotional distress that accompanies COVID-related losses. We must educate ourselves, listen more and talk less, and recognize our lack of expertise regarding the full weight of the virus. It is up to us to work through our own frustrations at feeling held hostage by a foe that didn't personally annihilate us, to shore up the gaps in our knowledge, and to lead with empathy for the profound loss of so many lives.

In truth, the horrific reality of losing loved ones without getting to say goodbye in person along with the permanent loss of physiological function—while half the world decries COVID-19 as a hoax—has taken a massive toll on millions of people worldwide. When we bear witness to all the stories, it is difficult to retain a lack of empathy or disbelief in the reality of the brutal effects of the SARS-CoV-2 virus. So, integrating the very real and very sad stories as we re-establish a new normal will be healthy for us—and for the whole world.

RESTART WITH EMPATHY

- **Take stock of feelings and friendships.** Have you faced significant conflicts, rejection, or disappointment in others in response to the pandemic? What are the resulting feelings? What can you do to work through these feelings and grieve the losses of relationships?
- **Recognize privilege if it exists.** Have you known people or communities who have faced trauma or loss as a result of the virus? If not, recognize your privilege and lack of expertise on the subject. Find ways of growing your empathy toward those whose experience is different from your own and respecting their experience with your words and behaviors.
- **Know more than one story.** Are you living from a "single story" that will profoundly impact the way you approach re-entry? If so, seek out stories from all sides of the issue. Embracing the complexity of many stories is crucial for personal and communal healing.

NAVIGATING DIFFERING NEEDS

None of us are undertaking re-entry in a bubble. Individuals who have been sheltering with others will need to consider their individual re-entry choices within the context of those with whom they live. Parents will face conflicts between their own needs and the needs of their children. Partners may face re-entry with entirely different needs and wishes. Extroverts may need to pay special attention to thorough communication and consent as they navigate their overeagerness to re-engage in in-person settings. Introverts may struggle to maintain solitude in a world hyper-attuned to efforts to reconnect. Employers will be required to balance the well-being of their employees with the needs of the marketplace.

Navigating differing needs will surely be tricky for all of us, and we're bound to make mistakes. In asserting our own needs, we'll likely make missteps, sometimes leading others to feel unsafe. Our ability to initiate and have complex conversations will be our best tool here. Conflict avoidance will make things worse. If one person's needs put others at risk, that person should not ignore those needs and resent the others, because doing so would lead to greater problems. Instead, being prepared to communicate about our needs with those who will be impacted by our actions and being ready to empathically consider their responses will be key. Applying inclusive responses and creative thinking will help. Similarly, flexibility in our openness to such conversations, initiated by others, will help us thrive in this time.

RESTART WITH CLEAR COMMUNICATION

- **Identify relationships that will need attention as you re-enter embodied life**. What relationships might require consent or communication before you move ahead with your own re-entry plan? In what ways do your personal needs in re-entry conflict with those of your housemates, coworkers, and community?
- **Prepare for difficult and/or complex conversations**. What is your comfort level with difficult conversations? If it is low, what is one small step you might take to help you with this? Consider practicing a conflict discussion with someone in role-playing mode.

AWKWARDNESS, HESITANCE, AND CONFUSION IN DETERMINING NEW SOCIAL NORMS

A pair of road signs, near the home of a friend, always catches my attention. The signs stand nearly eight feet apart, are rhombus shaped and yellow, and are positioned near a small park. The first one says, "DANGER," and the second one reads, "Playground." Before the pandemic, I thought these signs were hilarious. Now, after twelve-plus months of quarantine, they make absolute sense.

Many parts of our return to embodied living will be fraught with difficult decisions that will require critical thinking and consideration of things we never imagined would go together—like "Danger" and "Playground." Once someone is vaccinated, we wonder if it is acceptable to gather with others who are also vaccinated. Given evolving science, should we continue to keep children's play parallel and at a distance, or is it okay to resume playdates with less restriction? Is it safe to linger in a bookstore or hang out in a coffee shop, or should we continue to be hyper-vigilant about making our outings brief and limited? The kinds of mental gymnastics involved in our return to community life will be constant and complex. Just as we've experienced fatigue in the need to comply with strict guidelines, we'll face a new kind of fatigue determining what is and isn't acceptable behavior.

Although we'll face many of these decisions on our own, we also will need to negotiate the needs of our communities. Unfortunately, we'll be doing this from the position of relative isolation from them over the past year. This lack of connection will have negatively impacted our skills in negotiating social demands and situations. Such social deprivation and its impact on our communication and negotiation skills will have a bearing on how we navigate our independence and interdependence in this time.

According to the American Psychological Association (APA), *social deprivation* is defined as "limited access to society's resources" and a "lack of adequate opportunity for social experience."[2] For many of us, this very real experience has left us feeling bereft of new ideas or things to talk about with others. With our opportunities for social engagement limited, we lack the kinds of experiences that give us fodder for conversation topics or new ideas. We both desperately want to seek out connection and are absolutely nervous about doing so. This creates an unconscious conflict between the impulse to put ourselves out there in the world and the action it takes to actually do so. In this space, we can easily decide not to risk the effort required to do the difficult work of connecting. Craig Haney, professor of psychology, says, "[We have been] deprived of natural, normal, social interaction—all of these things that connect us to one another have all become problematic and prohibited." He likens this to other populations that have experienced social deprivation, such as formerly incarcerated individuals who, his research revealed, "end up self-isolating because they don't feel comfortable around other people because they've been isolated so much. When they have the opportunity to be around people, they don't take it. They feel awkward, pull back and keep to themselves."[3]

It's likely our early efforts to re-engage with our communities will be peppered with the kind of awkwardness that Haney describes. For individuals who have moved, experienced life-altering illnesses or accidents, or initiated or experienced other major changes (e.g., coming out, significant changes to their visual appearance, or the worsening of a chronic illness) during the time of quarantine, the awkwardness around sharing physical space with others may be overwhelming and filled with fear about how others will react. For those flooded with relief and anxious to connect as quickly as possible with people, the chance of people responding to them with hesitancy and boundaries seems high.

While many people are anxious to re-engage with their communities in embodied ways, the COVID-19–sanctioned opportunity to "keep to themselves" is exactly what many people would like to continue. Introverts who, from the start, emotionally and interpersonally benefitted by social-distancing guidelines may now face the need to find a way to "justify" maintaining significant chunks of solitude while much of the world wants nothing but connection. A wise individual, speaking to this reality, put it this way, "I'm dreading the pace of life speeding up and not knowing how to say no to all the things I've been pretending I've been missing." Maintaining these boundaries without feeling shamed will be difficult work.

As humans, we often don't relish situations that are unknown or could result in discomfort or awkwardness. It seems likely, however, that the bulk of our experiences in creating new social and personal norms will fall into these categories. We'd be well served by thinking these things through and practicing ways of tolerating the unknown and uncomfortable.

THE IMPACT OF REPETITION AND PRACTICE

While practice may not make us perfect, it certainly leads us to become proficient, and, during the pandemic, we practiced various behaviors that have altered the way we live. As stated earlier, in the early days of the pandemic, we were running a marathon but thought it was a sprint,

---

RESTART WITH COMMUNITY ENGAGEMENT

- **Take stock of how quarantine has helped you and/or hurt you socially.** In what ways have the global safety requirements of quarantine cost you? In what ways have they served you?
- **Prepare yourself for awkward social moments.** Practice self-soothing and affirming statements like, "Everyone deals with feeling awkward at some point or another" or "I am far more sensitive to my blushing than anyone else," and identify ways to take a time-out (in the bathroom or by stepping outside) when these moments occur. Keep in mind that it's okay to take time-outs and to practice prepared comments when you feel caught in a difficult social situation.
- **Internalize the reality that living through awkward moments is good for us.** Getting through difficult social situations builds grit and resilience. The goal should not be to avoid them but, rather, to find ways of helping ourselves through them and being gracious with ourselves as we do so.

---

and we hadn't trained for either. This led us to fall into many habits we thought we'd utilize for a few months and then "return to normal." Over time, however, these habits have become deeply entrenched and will not disappear overnight. Consider that runners who haven't trained for a marathon come to rely on "work-arounds," or habits that eventually can hurt them, if continued.

We'll discuss how habits are formed and broken in later chapters of this book, but, for now, let's consider some of the ways in which behaviors, repeated over the last many months, bring us to re-entry with many habits that may not be conducive to health and contentedness. While pandemic-related habits connected to what we wear, and do not wear, to work may be disappointing to change, those we've established related to our technology engagement, screen time, and social connection may feel downright impossible to break. If we don't address them, however, the long-term effects could be significant.

Long before the pandemic required us to maintain social distance, research told us that Americans were spending 10.5 to 12.5 hours per day in front of screens. This left us looking more outside ourselves for sources of connection, comfort, entertainment, and validation. The screen time also impacted our brains, leaving us bereft of robust wiring in the brain regions related to focus, delay, and emotional regulation. In addition, a reliance on text-based communication and social media platforms to keep us connected meant that our in-person and verbal communication skills were underutilized.

With digital worlds offering interesting, immersive, and perfectly curated spaces for education, work, entertainment, and connection—and embodied worlds without such robust novelty—we've come to rely on our devices in more profound ways than ever. When we finish one show, several others are offered that are right up our alley. When we run low on a pantry item, we can have it delivered at a single command. When we feel lonely, we can scroll through social media to get a sense of connectedness. These things have become only more ingrained during quarantine.[4]

The difficulty with such a reality lies in the fact that, while this reliance upon technology to meet our needs has been a lifesaver, it also comes with great costs. Very likely we are arriving at this time of re-entry feeling dysregulated and exhausted, unable to focus, and comparing ourselves with those we visit only in social media spaces. We feel vaguely or profoundly disconnected from our bodies and feel strange when we leave our homes. In our overexcited embrace of technology, we've ignored important experiences that help us develop grit, resilience, and a strong sense of self. The resulting vacuum will be felt in personal and communal ways.

We now roll out of bed and plop into our work or school chair with no need to tend to personal hygiene or travel. We overstimulate ourselves with our screens to avoid the boredom and lack of novelty at home. We are out of the habit of expressing ourselves verbally to others we casually encounter throughout the day. We are habituated to high levels of multitasking that will cost us to maintain as we reconfigure our workday to being elsewhere than home. We do the laundry between assignments, go on walks or work out during meetings, and tend to personal tasks during the business and school day. All these actions have created new rhythms to which we've grown accustomed.

Re-entering community and embodied life will require the awkward task of engaging skills and behaviors we haven't used in a while, and most of us don't relish putting ourselves in situations outside our comfort zones. We'll need to break some of our newly established habits, and we won't always want to. It's been convenient to work from home, and we've enjoyed the lack of commute. We feel disinclined to return to professional clothes, and we could do with never sitting through a boring meeting in person ever again. Students have had the benefit of Google assistance at all times in ways that won't be acceptable in classrooms and hesitate in losing this resource. Some of our habits, however, may be actively contributing to a lack of health and wellness. Pushing ourselves to move and use our bodies more consistently, to think more deeply and critically, requiring ourselves to get comfortable interacting with people in embodied ways, and re-establishing a connection to a wider world will force us to activate habit-breaking behaviors.

In addition to the way our new habits have passively shaped us, we must acknowledge that, for over a year, we've simultaneously been actively required to practice hesitancy and restraint in forging human connections with others. This habitual pause and evaluation in response to human interaction is likely to have an effect on our readiness and skill in re-engaging in-person community. Our brains and bodies don't just flip a switch and have a long-standing "forbidden" behavior suddenly become "okay." We may feel overanxious to re-engage with others in shared spaces, and we would do well to prepare ourselves for a certain level of discomfort and awkwardness that may accompany that re-entry. We also would benefit from considering the comfort and experience of those with whom we will be connecting. Approaching others with empathy and being diligent about asking for consent before moving closer than six feet, removing a mask, or reaching out to touch people will go a long way toward making re-entry smoother for everyone.

INDIVIDUAL, COMMUNAL, AND COMPLICATED GRIEF

If you asked me to name one emotion that almost everyone has experienced over the course of the pandemic, "grief" would be my response. Loss has been inherent in this time, and, whether we recognize it or not, we all have losses to grieve. Grief represents the physical, emotional, mental, social, behavioral, and spiritual reactions to any loss. Although many are familiar with Elisabeth Kübler-Ross's five stages of grief—

RESTART BY IDENTIFYING HABITS

- **Make a list of habits.** What habits have you established during quarantine? How have your daily routines changed? Have you developed new communication habits? Do you habitually google the answers to questions or conundrums before applying your own critical-thinking skills?
- **Evaluate the health (or lack thereof) of each habit.** Which of these habits are serving you and your physical and mental health? Which would you like to find a way to break?
- **Seek out resources for help with habits you'd like to break.** Many habits will be able to be addressed by the habit chapter later in this book. Others, however, such as substance use or habits related to mental or physical health, may need some additional tools. Find reliable sources to direct you to educational and support information, tools, and direct services.

---

denial, anger, bereavement, depression, acceptance[5]—it turns out that the process of working through loss is much less linear and prescribed than researchers originally thought.

Rather than the grief process seeming like a straight path, the experience is fluid and highly personal. Disorientation, strong emotions, and fluctuating moods are its hallmarks. There is no preset amount of time that is "typical" for completing the process, and it may linger throughout the course of a person's life.

Although grief is, possibly, most closely related to death and dying, it also relates to all kinds of loss. The loss of a given role in life (e.g., spouse, teammate, or vocation) can trigger grief. Grief also can be felt with the loss of important objects (e.g., items lost in a robbery or house fire) and experiences (e.g., not getting to participate in a senior year on campus). Likewise, losing an important vision of one's life can instigate mourning (e.g., experiencing a change in career or physical capabilities), as can the loss of security and safety (e.g., income loss, houselessness, and more). It is even possible to grieve over lost ways of thinking. This is the case for people who find themselves leaving political or religious ideologies that no longer fit or serve them. For many, this time of quarantine has offered ample opportunities for this kind of loss, as the assumptions and ideologies of being safe in our modern world have been challenged. When considering this list, it's easy to see how many types of grief have visited us during the pandemic and how intertwined our losses have been. If we have lost a partner to COVID-19, we've also lost the role of partner, the vision or dreams we had for the rest of our lives with them, and, possibly, income or other sources of security. The multiplicity of losses compounds and requires attention as we recover.

"Complicated grief" is a specific form of mourning highlighted by ruminating on the circumstances of the death or loss, worry about its consequences, and avoidance. Social worker Katherine Shear, MD, writes, "Unable to comprehend the finality and consequences of the loss, grieving individuals experiencing complicated grief resort to excessive avoidance of reminders of the loss as they are tossed helplessly on the waves of intense emotion."[6] The symptoms of complicated grief are preoccupation, pain in the same area as the deceased experienced, strong emotion in recalling memories, avoidance of reminders of the loss, feelings of emptiness and longing, preoccupation with thoughts of the loss, hearing the voice of the person who has died, and a feeling that it is unfair to be alive when others are not.[7]

Given the way the SARS-CoV-2 virus has impacted our daily life, it has been impossible not to be preoccupied with the losses we've experienced. Everywhere we look we are reminded of lost people, experiences, freedoms, ideologies. For those who have lost personal function or loved ones to COVID-19, this reality is even more intense. As a consequence, whereas in pre-pandemic times complicated grief impacted 7 to 10 percent of those in mourning, that percentage may be much higher now.

Research suggests that losses involving human error may be even more complicated to process than those that occur naturally.[8] Although trauma is inherent in many losses, understanding that the extent of losses could have been mitigated by earlier and more aggressive human intervention will make grieving COVID-19–related losses much more complex for many. Being able to consider this reality and work through the feelings related to it will be crucial in dealing fully and honestly with the grief. For many, this will be a highly uncomfortable part of the mourning process including anger, disappointment, and despair.

Another important point to ponder is that individual grief is impacted by its context. When entire communities are experiencing a shared loss, individuals can either feel invisible or understood and supported. Examining how our individual experiences of grief have been impacted by the global and local mourning related to the pandemic will help us navigate our personal grief processes and heal.

RESTART WITH GRIEF

- **List those within your sphere who have died or who have had loved ones die of COVID-19**. Identify any and all lives lost to COVID-19 in your personal sphere. If you can't name anyone personally, remember that this is a rarity and that your story is not the primary narrative. In this case, seek out stories of those who are grieving loved ones lost to the virus.
- **If you've had COVID-19, list the symptoms and losses that you've experienced**. Take some time to acknowledge these as well as any remaining symptoms. Make space to recognize these as very real and understandable losses to be grieved.
- **List the experiences that you've missed out on during quarantine**. This could include graduations, trips, retirement celebrations, birthdays, cultural events, and large personal and family milestones. If you've lost someone to COVID-19, this might include in-person memorials or funerals as well as the experience of getting to have said goodbye in person.
- **List any other kinds of losses triggered by the pandemic**. These could include economic losses, loss of important roles or relationships in your life, or ideologies.
- **Offer yourself a way of memorializing these losses**. Find a way to honor and mark your losses. You could write them on pieces of paper and offer some space for feeling, and then burn them to release it all or create some other form of ceremony.

PERSONAL TRAUMA AND COLLECTIVE DISTRESS

Trauma has become a word often used without a clear definition. The Substance Abuse and Mental Health Services Administration (SAM-

HSA), however, defines it this way: "[T]rauma results from an event, series of events, or set of circumstances that is experienced by an individual as physically or emotionally harmful or life threatening and that has lasting adverse effects on the individual's functioning and mental, physical, social, emotional, or spiritual well-being."[9] Trauma is stored not only in our minds and emotional cores but is also encoded in our cells and is carried forward from the traumatic event. Rather than simply staying attached to the instigating experience, it can be triggered at any time, leaving us feeling overwhelmed or hyper-vigilant, fed by activation of the fight/flight/freeze/faint response region of the brain.

It's fair to say that many people around the globe have experienced trauma over the course of the pandemic. For those who have contracted the virus or lost loved ones to it, the trauma is likely significant. The same is true for frontline workers from medical staff and mortuary workers to essential workers such as grocery store staff. Even for the general public who takes seriously the life-threatening nature of COVID-19, however, there likely have been experiences during the pandemic that the mind and body have encoded as trauma. The emotions and physiological experiences that go along with trauma will, in these cases, be piqued with every mention or confrontation of the virus's ongoing impact in the world.

In general, some people are more likely to experience trauma than others. The same is true of Post-Traumatic Stress Disorder (PTSD). Mitigating factors include prior exposure to high levels of stress and being part of a marginalized (or bullied) people group. The way we process stress also can impact the way we do, or do not, experience traumatic responses. "[The] drastic differences in coping with stress, anxiety, and trauma appear to have both a biological and environmental basis," reports Erika Hayasaki in her feature article about Margaret McKinnon, a traumatic event survivor and leading researcher on how stress is stored in the body and brain. Hayasaki continues, "For most people, a sudden car accident, an imminent plane crash, a life-threatening attack, or a brush with someone who might be infected with a novel virus can kick up the fight-or-flight response, releasing hormones such as cortisol and epinephrine that propel energy to muscles. Neurotransmitters such as norepinephrine, adrenaline, and dopamine filter into the amygdala, stimulating the brain to tell the heart and lungs to beat and breathe faster.

Emotions and acuity go on high alert."[10] While these reactions commonly accompany distressing experiences, not everyone will develop a full trauma response.

In sum, the same experience can result in very different outcomes for individuals. Whereas many individuals will be re-entering the world with COVID-19–related trauma, others will be going on their merry way, oblivious to how their words and actions may trigger another person. This is a setup for heightened personal harm and cultural conflict. For this reason, everyone would benefit by adopting a trauma-informed stance while reconnecting with others. Being trauma-informed simply means we are aware that the experience of others is different from our own and may be woven through with the lasting emotional, intellectual, behavioral, and spiritual effects of having encountered a traumatic event or series of events.

Activist Ashley Chec offers the acronym "CARE" to help people lead from a caring and trauma-informed stance. By leading with Clear and authentic communication that is based in respect for self and other and the autonomy of both, we set the stage for people to bring their honest selves to encounters rather than to try to hide or hyper-attend to their trauma. Once the foundation for this kind of interaction is established, Active, empathic, openhearted, and open-minded listening will allow us to learn and the people experiencing trauma to express what they need without fear of being shamed or shut down. At root, these behaviors will communicate Respect for the autonomy and choice of those experiencing trauma. People will feel seen and heard if we are able to respect their experiences, even if they differ from our own; living life from such a stance will create safe space. When all this is done with care and intention, the person with trauma will feel Empowered to keep showing up, and we also will feel empowered to continue to offer empathic spaces for those who suffer.[11] It would do us all well to adopt this approach as we re-enter the world.

For those individuals living with trauma and/or PTSD, talk therapies as well as some psychotropic medications can be helpful. EMDR (Eye Movement Desensitization and Reprocessing), a form of therapy involving recalling a traumatic incident and following guided eye movements that assist in rewiring the brain, is also highly effective. TMS (Transcranial Magnetic Stimulation) also shows promising benefits for those living with PTSD, and early research suggests that therapies involving guided use of psychedelics may prove beneficial for the treatment of PTSD.

RESTART WITH TRAUMA

- **Check your interpretation.** Does the word *trauma* invoke a response in you? Do you have a bias against people who have experienced traumas of different sorts? Have you experienced trauma but resist naming it as such?
- **Consider your experience with a trauma lens.** Looking at the SAMHSA definition stated at the beginning of this section, do you think "trauma" accurately describes any of your experiences during the pandemic?
- **Practice a non-judgmental pause and response.** If hearing others refer to their own trauma or difficulties sparks judgment and/or irritation in you, practice noticing that feeling and linking it with a pause before you internally judge them or externally respond with impaired empathy. In this pause, practice mantras like, "Everyone has their own journey, and I don't need to weigh in on theirs" and "Empathy is the goal, not agreement."
- **Take action.** If you have experienced trauma, consider taking at least one action to better understand or work through your experience. Read articles about it on the SAMHSA or American Psychological Association websites or make an appointment with a licensed mental health professional who has expertise with trauma.

LONELINESS

Prior to SARS-CoV-2 hitting the country, a growing loneliness pandemic existed in America. Popular press outlets ran feature articles about it, and social media giants made attempts to understand their roles by assembling research groups and think tanks. It was becoming clear that our reliance upon the more fleeting forms of connection offered in digital domains was not meeting our actual needs for belonging. In 2017, researchers found that two in five Americans felt their relationships lacked meaning, and one in five reported loneliness and feelings of isolation.[12]

Loneliness impacts both our physical and mental health and has been shown to be as damaging to health as smoking fifteen cigarettes a day.[13] Research further demonstrates that poor social relationships are associated with a 29 percent increase in risk of coronary heart disease and a 32 percent rise in the risk of stroke.[14] In 2019, loneliness was widely identified as a public health issue, impacting, in particular, millennials and aging boomers. Media outlets from *Forbes*[15] to *Scientific American*[16] covered the topic. It's clear the physiological and psychological impacts of loneliness are real and pressing—and there's little doubt the pandemic has intensified these realities.

The news, however, isn't all bad, as research studies in the United States and United Kingdom both show a leveling of the rate of loneliness during sheltering-in-place orders.[17,18] A journalist covering the research for *Scientific American* explained, saying, "Surprisingly, it seems that levels of loneliness around the world have remained generally stable [during quarantine]. How have we avoided a social fallout? First, social isolation does not necessarily cause loneliness. While isolation is the objective state of being alone, loneliness is the subjective experience of disconnection, which means that you can feel lonely while surrounded by people or connected while by yourself. Amid COVID-19, most of us are more isolated, yet that doesn't mean we are lonelier. . . . In the past few months, we've made a point to prioritize connection. The pandemic has made people more aware and appreciative of their relationships."[19]

It appears the pandemic may have led many of us to find novel ways of connecting with others. Whether this was through shared space in video game environments, on Zoom or other kinds of video or audio calls, or through watch parties or shared online experiences, we discovered ways to be together and to feel connected. In addition, new volunteer opportunities arose for us to engage with a wider community. These choices made it such that many people actually experienced less loneliness during the season of sheltering than they did in previous seasons of life. In many ways, technology became a great equalizer for some, offering opportunity for connection regardless of geographic location, thus expanding social circles.

This is not true, however, for everyone. For those sheltering entirely alone or with particularly large risk factors, uncomfortable in the digital domain, lacking reliable access to it, or particularly desirous of the energetic exchange inherent in embodied connection, feelings of isolation and loneliness have grown since quarantine. For these people, the ability to maintain connection both with individuals and current events may have taken a real hit. While others have grown their proficiency and creativity around staying connected, these demographics have been left out and are at risk of feeling both lonely and forgotten or abandoned. Feelings of incompetence or a hyper-awareness of what they lack—in the way of ability, skill, literal hardware, or potential—mix with the experience of isolation to make these people's experience of the pandemic particularly tricky.

As we migrate parts of our relational lives back to shared physical spaces, everyone will experience differing levels of relational complexity. Those who have experienced growth in their interpersonal connectedness through digital communities and experiences may feel abandoned as others depart digital spaces for embodied ones. Those who experience difficulties in expressing themselves in physically proximal spaces will have special adjustments to make, having been made comfortable by the global move to screen-based connection and social distancing. For those who have experienced real isolation and the diminishment of their social support system, re-entry may be met with fear, given the isolation and lack of practice inherent in the past year.

RESTART WITH AN EXPLORATION OF LONELINESS

- **Get clear about what you are experiencing**. Distinguish between feelings of loneliness and feelings of anxiety over lack of practice with social gatherings. If you're anxious, begin practicing social interaction in small ways at the grocery store or coffee shop. If you are lonely, work to identify what you are lonely for specifically. Conversation? Sharing experiences with someone? Get as specific as you can and make plans for taking steps toward getting what you need.
- **Work to identify possible relationships to build**. Consider your values and interests and identify volunteer or social opportunities that might align with them as well as offer you access to new people. As the world re-opens, find classes or experiences to engage that will put you in the way of others who share your interests. Practice a conversation starter or two that you can use to initiate conversation.

PHYSIOLOGICAL HEALTH ISSUES
MADE WORSE BY PANDEMIC STRESS

Fear and stress have a profound impact on our physical health, and the last year has been filled with both emotions. Dr. Judith Cuneo, associate director of clinical programs and an integrative obstetrician-gynecologist at the University of California San Francisco (UCSF) Osher Center for Integrative Medicine, explains this relationship between distress and health well. She writes, "Just a low or depressed mood, simply sadness, can affect people's resilience, and over time that sort of chronic stress and low-level stress can begin to affect our health and reduce our immune function and worsen chronic diseases that are present and stress-related illnesses that are present." In essence, prolonged stress can instigate a domino effect that eventually impacts our immune systems and cardio-vascular health and worsens chronic pain and other conditions. Dr. Cuneo notes, "[L]ow-level chronic stress can put us into that sympathetic nervous system activation, that 'fight-or-flight' stress reaction. What follows is a cascade that happens that we may not be aware of, physical changes like our heart rate [go] up, our blood pressure goes up, our hyperarousal deteriorates."[20]

We have all been living in stressful conditions. Further, with so little novelty in our lives, many of us may have become hyper-attuned to our bodies, experiencing anxiety with each noticed sensation, while others have been in denial, focusing instead on getting through by whatever means necessary and overlooking physical concerns. Complicating things, many people have gone without routine medical care during this time. Those who entered quarantine with preexisting physiological ailments will need to get caught up on missed physician visits and interventions as well as to develop practices that help release stress and powerful emotions. For all of us, taking stock of our physical health and examining any way in which our stress has impacted our bodies is important. Paying special attention to sleep, nutrition, and exercise as well as to any mental health concerns, it would do us well to notice changes.

Mindfulness-Based Stress Reduction (MBSR) is an especially helpful tool for mitigating the impact of stress on the body. Instructors of MBSR offer evidence-based programs that teach people how to use meditation as a tool for managing stress, depression, anxiety, and pain.

Developed by Jon Kabat-Zinn and the University of Massachusetts Medical Center, MBSR encourages non-judgmental acceptance and investigation of thoughts, feelings, bodily sensations, and present reality to increase well-being.[21] Many therapists and practitioners around the world teach MBSR skills, and there are also comprehensive online instructional programs; engaging their expertise would be particularly wise at this point in time.

RESTART WITH PHYSICAL HEALTH

- **Take stock of your health.** What routine checkups or doctor visits have you skipped because of quarantine? Are there any physical symptoms that you have ignored or been hyper-aware of?
- **Take action on catching up.** Set aside an hour on your calendar to schedule all of these appointments.
- **Try MBSR.** Find a local teacher or look online to find the description of MBSR practices. Commit to learning one MBSR practice and doing it daily for a week. Take note of effects.

MENTAL HEALTH NEEDS

The following sections address the major mental health needs that have arisen during the time of quarantine. I'd like to begin by establishing some context for the information contained therein.

The American Psychological Association has conducted "Stress in America" surveys throughout the time of quarantine. Around the pandemic's one-year mark, the findings revealed all-time highs in people's self-reporting of mental health concerns. In the February 2021 survey, 84 percent of adults reported experiencing at least one emotion tied to prolonged stress in the prior two weeks. The most common emotion was anxiety at 47 percent, followed closely by sadness at 44 percent and anger at 39 percent.[22] These numbers are reflected in the fact that, even a few short months into quarantine, calls to crisis lines hit new highs across the United States—and the psychological toll has not lessened.

The kinds of resources available to those living with mental health difficulties have always varied by geography, as well as by monetary and personal privilege, and are often complicated to access. Long before the pandemic, many groups such as children, people of color, those living in poverty, and those living without insurance remained underserved while, at the same time, stigma about seeking mental health help kept many others from finding the care that could benefit them. This problem has only been made worse during quarantine, which both moved services online and overtaxed the mental health system. Although it would seem as though the mass move to telehealth might have made services easier to access, the reality is that there are far fewer providers in the United States than there are those in need of care. As a result, large numbers of people are currently suffering without the help they need.

Massive numbers of humans are currently living with depression, anxiety, and substance dependence that has developed or worsened over the course of the COVID-19 pandemic. Once these, or other mental health difficulties, settle in, they don't go away overnight. This means that the impact of increased psychological distress may be felt by individuals and communities for a long time. Left untreated, complications arise in cascading ways. Depression and anxiety can lead to substance misuse or suicidal ideation or plans. Substance misuse can enhance relational difficulties, which can intensify symptoms and vice versa. We need to take seriously the presence of any mental health concerns and address them proactively as individuals and communities. To that end, the four most common mental health concerns to intensify during COVID are explained briefly here to inform and empower the search for appropriate help and interventions.

ANXIETY AND DEPRESSION

"To be deprived of contact with other people, to be deprived of natural, normal, social interaction—all of these things that connect us to one another have all become problematic and prohibited," says psychology professor Craig Haney. He continues, "And this is producing depression, it's producing anxiety, it's producing a destabilization of their sense of self."[23] According to the APA, people with depression may "experience a lack of interest and pleasure in daily activities, significant weight loss or gain, insomnia or excessive sleeping, lack of energy, inability to concentrate, feelings of worthlessness or excessive guilt, and recurrent thoughts of death or suicide."[24] When individuals have sustained symptoms that interrupt their ability to function at "normal" (for them) levels for two weeks or more, clinical depression is likely in place. Thankfully, treatments including talk therapy and psychotropic medications, as well as some evidence-based alternative treatments, work well for helping people heal from depression's painful effects.

The risk for depression is higher for some groups at this time. Individuals who were impacted by the economic downturn, as well as communities of color, and children and their parents (specifically mothers) all appear to be at greater risk. In addition, essential workers and those who have lost loved ones to the virus run the risk of experiencing higher levels of depression than the general public. Those who lived with depression prior to the pandemic may also be at great risk for a severe worsening of symptoms throughout quarantine. Special attention should be paid to these groups.[25]

Anxiety is defined by the APA as "an emotion characterized by feelings of tension, worried thoughts, and physical changes like increased blood pressure. People with anxiety disorders usually have recurring intrusive thoughts or concerns. They may avoid certain situations out of worry. They may also have physical symptoms such as sweating, trembling, dizziness, or a rapid heartbeat."[26] Panic disorders fall under the category of anxiety as do phobias. As with depression, experiencing a mix of these symptoms for two weeks or more at a level that interrupts normal levels of functioning may mean a diagnosis of anxiety is in order. There are several gold-standard treatments to help those who live with anxiety disorders: Cognitive Behavioral Therapy (CBT) as well as Acceptance and Commitment Therapy (ACT) are particularly effective. Medications can help, too.

Adolescents and young people have experienced unique peaks in their levels of anxiety in this time, putting them at particular risk of long-term suffering if treatment is unavailable.[27] Social anxiety—characterized by persistent fear of single or multiple social situations or experiences wherein one must perform, exposing them to unfamiliar people or to possible scrutiny by others—may have been especially intensified as a result of the social-distancing requirements related to the pandemic. The pandemic has wreaked havoc on our anxiety levels. Untreated, it will continue to have a profound effect on health as well as on our ability to maintain healthy relationships and function with ease in the world. Finding professionals uniquely capable to treat anxiety in all its forms will be crucial for people who are suffering with it.

SUBSTANCE USE

Mental health researchers and clinicians have noticed an uptick in both quantity and frequency of substance use during the pandemic. Not only did drug-related overdoses spike during this time, but multiple disciplines also report increases in the use of opioids and stimulants across the country.[28] Although these increases cannot be definitively explained by the pandemic alone (for instance, availability of drugs must be considered), it seems clear that the economic instability, the fear of contracting the virus, the social isolation, and the life disruption inherent in this time have led many to self-medicate. The reality that many people are using substances in solitude might also explain some of the increases in reliance upon substances for coping as well as the possible lethality of use; when using drugs alone, no one is present to intervene in the case of a serious reaction.[29]

Medical and psychological interventions are important for helping us dial back our reliance upon substances. Because of the migration to telehealth services and the opportunity this has provided for medical professionals to prescribe medicinal assistance without a physical examination, access to treatment is much easier (even for those who misuse drugs). For those lacking access to reliable technology or internet, however, treatment may be difficult to receive. Being vulnerable about asking or offering help will be key to overcoming these challenges. We must be bold about removing the stigma and shame around seeking help for substance use.

DOMESTIC VIOLENCE

The increase of intimate-partner violence and child abuse since the beginning of quarantine has recently been referred to as "the pandemic within the pandemic."[30] Increased stress levels are often correlated with increases in abuse, and isolation almost always makes stress worse. Violence within relationships has intensified in situations in which it existed before the pandemic, and it has been initiated in others.

Stay-at-home orders have had particularly devastating effects for those sheltering with individuals who perpetuate violence. The social isolation, closing of schools and workplaces, as well as the rise in mental health concerns means that many abuses have gone unnoticed outside the home. The shame that often accompanies domestic violence has always made it difficult to seek help, and this is also made worse by the lack of privacy in shared quarters in this time and the reduction in available services during the pandemic.

Data from US crime reports show that one in six homicides are related to intimate-partner violence and that those subjected to it are at greater risk for long-term emotional and physical difficulties of all kinds. For children who have experienced abuse at the hands of caregivers or witnessed it between caregivers, the risk of lifelong health and emotional consequences is particularly high.[31] Because of these stark realities, we all must be ready with information about shelters, therapists, and community agencies for those who need them. Educating ourselves about how to have difficult conversations in non-shaming ways, while inviting those who have suffered to be honest about their situations, will also go a long way toward helping people leave abusive homes and find safety.

OTHER PSYCHOLOGICAL IMPACTS

In addition to depression, anxiety, substance misuse, and increases in domestic violence, a few other important mental health trends are worth mentioning. Specifically, the impact of a sedentary lifestyle on emotional well-being, the sleep disturbances brought on by the pandemic, and the reality of emotional burnout are all very real concerns. Also, individuals living with mental health concerns exacerbated by isolation are likely suffering with greater symptoms than they may have had pre-pandemic.[32]

RESTART WITH MENTAL HEALTH

- **Collect resources to have on hand when you or someone you love need them**. Search for mental health professionals and resources such as substance use groups and support organizations, community centers with counselors for children and youth, as well as domestic violence resources and shelters. The resource list in this book will help.
- **Prepare yourself for how to talk with someone about mental health concerns**. Seek educational resources for how to have difficult conversations about mental health so you can be ready to broach the subject with someone when the opportunity arises. People living with mental health concerns are usually relieved by invitations to open up about their issues with others who are safe and receptive.
- **Seek out high-quality help**. If you are suffering from symptoms of a mental health condition and don't yet have a therapist, ask your friends, physician, faith leader, or other trusted individuals if they can connect you with reliable and trustworthy mental health professionals. As much as possible, be open to all potentially helpful interventions. Psychologists (PhD and PsyD), licensed mental health counselors (LMHC), licensed professional counselors (LPC), and licensed clinical social workers (LCSW) can provide talk therapy and other alternative supports. Psychiatric mental health nurse practitioners (PMHNP) and psychiatrists (MD) can provide medication and, in many cases, talk therapy. Many other "alternative" therapies are emerging as helpful tools in treating the difficulties mentioned here. Talk about the relevant options with trusted professionals. Finding the treatment most effective for you may take some time, but it'll be well worth the time and effort.

II

RESTARTING FOR
SPECIFIC POPULATIONS

3

FOR THOSE WHOSE LIVES
HAVE BEEN PERSONALLY
TOUCHED BY COVID-19

Just as a stone thrown into a calm lake creates concentric circles in the water at the surface, so will the impact of the virus be felt in waves emanating out from those whose lives it has touched. COVID has changed us all. We've lost a sense of innocence in regards to the way in which we move about the world, and our trust in each other has been deeply altered by witnessing a gobsmacking percentage of the population respond in unconscionable ways.

Several months into quarantine, I was invited to help facilitate weekly support groups with COVID Survivors for Change. While leading my second week of groups, the parents of my dear friend Kim were hospitalized with COVID-19. Within a week, both of them died. Kim sent pictures of her and her siblings trying to communicate with them, first, outside hospital windows and then later, with gloved hands and hazmat suits, saying their final goodbyes. These experiences have offered me a front-row seat to an experience that never should have happened. The unique trauma and grief that millions of people have experienced over the course of the pandemic is palpable and excruciating. We would all do well to read this chapter in order to understand the reality that many of our friends and neighbors face.

There are several broad categories of people for whom this chapter is specifically written. These people will be referred to as survivors. First there are those who have had the virus and recovered. Some who fit in this category experienced relatively minor symptoms and recovered fully with very little upset to their lives or bodies. Others, however, face the complexities of long-term physical symptoms, which include fatigue, cough, shortness of breath, joint and chest pain and escalate from there. Many "long haulers"—people who have had more than one episode of

COVID-19 or who never fully recover—experience inflammation of the heart muscle, lung abnormalities, acute kidney injury, and cognitive and mental health impairment.[1] For these people, unique challenges exist both physically and psychologically as the relative newness of the virus leaves them largely bereft of gold-standard treatments. For yet other survivors who may have been asymptomatic, living with the reality or fear of having exposed others to the virus, some of whom have died, fills their days with regret.

Next come those who have lived through caring for someone with the virus or who have lost a significant other (or others) to it. This is a highly diverse group of people who have dealt with an unknown trajectory of viral impact, an overtaxed health system, and/or death in very trying times. For those whose partners, parents, children, friends, or others were ill with, or died of, the virus in the early days, their experience is tinged with high levels of panic and regret. We knew so little at that time, and hindsight convinces many of these individuals that they could or should have done things differently. They will forever live with the ringing of sirens in their ears, the memory of rushed phone calls with physicians, and video visits with their loved one, cut short because of limited technology in the hospital setting. The feelings of isolation, stigma, and not knowing what to do enshrouded their traumatic experience.

Everyone in this broad category has likely experienced the complexities that have come with trying to help direct treatment and/or say goodbye with little to no physical access or proximity. There have been no hospital visits, no in-person physician consults, funerals were delayed or completely disregarded, and many have grieved without any family or friends nearby. In a high proportion of cases, the time between their loved one's diagnosis and death was very short, adding to the surreal feelings around the deaths, and many have lost more than one person, complicating things significantly.

Unique to COVID-19 survivors, this group has, at the same time, been faced with widespread societal denial of the severity of the virus, creating a horrible mix of feelings including anger, disillusionment, invisibility, and angst. This is made worse by being forced to live in a time rife with constant reminders of the virus that killed their loved ones. These people have been questioned by their family and friends (e.g., "They had a preexisting condition, right?") and have often had their losses seemingly dismissed. The corporate trauma that this group experiences is particular and painful.

Third are the professionals who have put themselves in harm's way to care for those with COVID-19. Next to the people who contracted the virus and their families, those working in health care, emergency response, and mortuary services had their lives forever changed in nearly an instant. Bereft of the quantity of Personal Protective Equipment (PPE) that a pandemic would demand and a centralized plan for how to slow the spread, hospital administrators, nurses, doctors, EMTs, sanitation workers, and all other manner of professionals swung into action. In the early days, in particular, the spaces within which they worked buzzed with panic, fear, anxiety, and death.

Burnout was, and continues to be, real for these professionals, who also faced ever-mounting scrutiny from the general public. They were required to take on tasks they'd never tackled while, at the same time, providing the sole emotional support to patients who were suffering alone. Cleaning professionals, supply deliverers, and others like them faced special stresses as they were repeatedly exposed to the virus without the commensurate PPE, praise, and kind of thanks offered to medical staff. The long-term impacts of the repeated trauma and prolonged stress that these professionals have faced are largely unknown. For those that have lived this time out in direct service, the needs for care and intervention will be great.

Every one of these groups will respond to re-entry in very personal ways. A commonality, however, will be a state of hyper-awareness around mentions of the pandemic and/or the unique way in which the virus has touched their lives. While those not personally impacted can hear news stories or chat casually about the frustrations of having freedoms limited, those who have been personally touched will likely be triggered every time COVID-19 comes up.

We are all going to need massive amounts of empathy as we move out and into the world again as we will never know for certain the kind of journey each of us is embarking on. There's no way to know who is a survivor and who is not, and to dismiss or deny the experiences of the more than 28,000,000 people whose lives have been directly touched by the virus would be akin to denying the presence of gravity. To help us fuel our empathy, and to offer some direction to those who have had, or lost loves to, the virus, the following are points of commonality to be addressed by all COVID-19 survivors and acknowledged by those who live with or near them.

THE UNIQUE PAIN AND ANGER OF HAVING PEOPLE DENY THE REALITY OF A VERY REAL VIRUS

When talking with COVID survivors, it doesn't take long to hear about the cognitive dissonance they bear witness to in their communities. They live with a particularly acute awareness of the brutality of the virus, while being confronted by a large segment of culture that doubts and denies its very existence. They encounter people who tout COVID-19 as a hoax and others who live as though it simply can't touch them. Daily, they encounter people who don't wear masks and learn of large gatherings that their friends have attended. Their senses are assaulted by the seemingly casual response of so many to something that has profoundly changed their lives.

The mix of anger, confusion, and hurt that this creates is unique and palpable. We can all imagine our lives being horrifically altered by a force that our friends and family fail to recognize as real. Over time, this kind of widespread denial, sitting alongside the vulnerability inherent to COVID-19–related loss, creates a set of strong emotions for survivors. Unaddressed, anger coupled with a feeling of powerlessness can lead to a particularly toxic internal state. This particular potent mix often leaves survivors on one or the other end of the "on edge/about to completely lose it" to "hopeless/despair/can't move" continuum. When we feel ourselves land on the continuum, or swing wildly along it, it's crucial that we step back and get some help.

FACING THE SKEPTICISM OF OTHERS AROUND YOUR ILLNESS OR THE ILLNESS/DEATH OF A LOVED ONE

Disenfranchised grief refers to any grief that goes unacknowledged and/ or is minimized or misunderstood by others. This type of grief can drive survivors to feel deeply unsupported and alone in their pain and is a par-

OWNING OUR FEELINGS AND
ASKING FOR WHAT WE NEED

- **Identify all the big feelings.** Until we've named all the big feelings that we've experienced (and are experiencing), we remain unable to ask for what we need.
- **Find supports and help with healing.** Emotions aren't rational so we're going to need some places to do the messy work of exploring them honestly in order to heal. Support groups, individual and group therapy, and having a safe set of people to talk honestly with can all help with this. Organizations working specifically with COVID-19 survivors are listed in the resources chapter of this book.
- **Inform others about what you need in the way of listening and support.** In searching for places to put and work through our experiences, it's crucial to remind anyone willing to listen that we aren't in need of being "fixed." Instead, we are in need of being heard and of having our pain and losses recognized. It's okay to ask them to hold off on offering advice or correction until we ask for it, directing them, instead, to simply listen and help us hold our feelings.
- **Find places to channel your big feelings.** Finding a COVID-19– or grief-related support group, continuing in therapy, or getting involved in making legislative or social change are great tools for this. Volunteering with COVID-19 relief programs may also be an empowering act.
- **For non-survivors . . . listen to survivor stories with these skills.** For those of us reading this chapter as a way of expanding our empathy, commit to being a good listener to stories of survivors. Withhold judgment (don't ask about preexisting conditions, etc.) and don't offer ways of calming emotions before simply bearing witness to them. The survivor is the expert here; learn from them and ask them what they need. Don't offer anything unless you will follow through. This group of people has lived with a lot of failed promises. Don't add to them.

ticularly potent shared experience for many COVID-19 survivors. The stigma and politicization associated with the virus, alongside the disillusionment that comes from observing the bold denial of its existence by people they once felt close to, intensifies the sense of isolation and suffering that many survivors share.

This isn't all that they have in common. There are also unique responses that survivors hear when they disclose their stories to people whose lives haven't been directly touched by the virus. When a survivor shares about the diagnosis or COVID-19–related death of a loved one, it is not at all uncommon to hear one of the following responses. "How old was (insert name)?" "Did they have preexisting conditions?" "Were they careful? Did they wear a mask?" "How did they contract the virus?" "Are you sure they actually died of COVID? You know, I hear they're putting that on a lot of death certificates even though the death was actually from organ failure/whatever other condition" or "You know, they were pretty overweight."

Grief expert David Kessler says, "Each person's grief is as unique as their fingerprint. But what everyone has in common is that no matter how they grieve, they share a need for their grief to be witnessed. That doesn't mean needing someone to try to lessen it or reframe it for them. The need is for someone to be fully present to the magnitude of their loss without trying to point out the silver lining."[2] With COVID-19–related loss, the risk isn't so much that someone will point out a silver lining but is, instead, that they will (often unconsciously) try to assuage their own fear. Basically, we are more than ready to minimize and attempt to compartmentalize the grief that survivors are experiencing because of our own fear that COVID-19 could touch us. By centering our response in our own questions rather than in offering empathy and bearing witness to survivors' losses, we forego having to be touched by the suffering that COVID-19 leaves in its wake. We feel more powerful when we can make sense of something (find fault in those who contracted the virus), so we seek reasons to explain why someone would die rather than seeking to bear full witness to their loved one's grief.

While there have always been ways of minimizing a person's grief with trite comments ("God must have needed another angel" "At least they had a long and happy life"), the statements constantly lobbed at COVID-19 survivors have a particular sting. They politicize personal losses, minimize the importance of their loved ones' lives, and call one's own integrity and value into question at a time when none of these responses is appropriate or called for.

STANDING UP FOR OURSELVES
IN THE FACE OF SKEPTICISM

- **Practice a strong response to questions**. Develop a standard response to off-putting questions. Find what feels comfortable and try to work it down to a sentence or two. Remember that it's okay to be emotional when you are facing scrutiny and to set boundaries about what you will and won't answer. Here are some starting points:
  - "Age and preexisting conditions are really irrelevant. My parent/partner is gone. I'm grieving. My loss feels belittled when people try to explain it away."
  - "Actually, those questions really hurt. My friend/grandparent had COVID and died. I need the emphasis to stay on their death and my loss."
  - "I feel very reactive when these questions are asked. What I need in sharing my child's/aunt's passing with you is empathy."
  - "I know that everyone is afraid of the virus so they want to try to understand who's getting it and how. In reality, though, these questions feel really dismissive of the loss I'm facing. Can we start over?"

  It may feel very scary to offer these responses but, with practice, having them in our back pockets can feel empowering, helping us know that we are prepared to care for ourselves and to honor our loved ones.

- **For non-survivors . . . respond to survivor stories with these statements**. For those reading this chapter to build empathy, consider sharing this information with your circles, encouraging people to offer compassion, rather than questions, to those facing grief. Commit to greater time spent listening than talking and practice responses that don't center yourself and your curiosity. Options are:
  - "I have so much compassion for what you must be going through."
  - "Grief is excruciating. I bear witness to yours."
  - "I can't imagine what you are going through. Is there anything that I can help you hold?"

HAVING SUFFERED (OR WATCHED
A LOVED ONE SUFFER) ALONE

There is a great deal of shared sadness and distress around the reality of people having suffered alone in this time. Whether virus-related isolation occurred at home (due to exposure or diagnosis without symptoms requiring inpatient care) or in the hospital, knowing that your loved one was/is handling the complexities of COVID-19 completely on their own is weighty. Add to this the lack of information about the virus and its treatment in its inception and the constantly evolving data as it spread, and it's easy to understand how frightened and hyper-vigilant we all were when we knew someone might have it. Knowing that they would need to go without our physical presence for support or intervention took a heavy emotional toll.

When speaking with survivors, it's not uncommon to hear them doubt themselves in relation to their interactions with physicians and nurses during their loved one's hospitalization or to regret not having "gotten more involved" in their loved one's care. Things were happening quickly and hospital staff had little time to communicate thoroughly. Patients didn't have access to technology that could connect them to their families outside, and reliable information about the virus or its treatment was hard to find and constantly changing. The regret-filled sadness of not having physical access to their loved one and/or those treating them is lingering and powerful. Feeling confident that one has offered comfort and care to those suffering helps people come to terms with the death. This confidence was completely unavailable to survivors.

Managing important conversations and processing loss through windows or over the phone left all parties feeling oddly distanced in what would normally be a very connected time. Deep regret over things unsaid, sadness that final goodbyes were offered only at a distance, and having not felt capable of being present when their loved one was taken off life support (either due to personal emotional limitations or hospital restrictions) are understandable experiences that survivors feel the weight of. While we all have the sense that everyone was doing the best that they could amid a very difficult set of situations, we also wish we could have known then what we know now. We wish our experiences could have been different. We wish, we wish, we wish.

Frontline workers have suffered from many of these same realities. Racing in to treat a completely unknown virus bereft of adequate information, research, personal protective equipment, and hospital beds, these professionals also suffered. Bearing witness to the struggles of their patients while fearing for their own health, working double shifts, and having to find entirely new ways of providing care took a heavy emotional and physical toll. Add to this the massive number of deaths that they attended, often as the sole support person to the dying, and we realize the trauma that these individuals live with. These professionals, wish, just like the families of those who have suffered, that they knew then what they know now. They wonder if they did everything they could and may live with regret. They wish they could have been more effective. They wish they could have shut the whole virus down. They wish, they wish, they wish.

DEALING WITH REGRET AND UNMET LONGING

- **Acknowledge regret and unmet longings**. Voicing them to at least one safe person who can listen intently will go a long way toward helping survivors and frontline workers make their way through feelings of guilt and regret. Finding others who have faced similar struggles can also be helpful.
- **Recognize when it's time for professional help**. If the weight of the experiences you've gone through is becoming increasingly difficult to bear, if you are experiencing symptoms of PTSD (intrusive memories, flashbacks, nightmares, hyper-startle response, and/or intense anxiety that disrupts daily life) or depression (increased guilt, lethargy, too much or too little sleeping and eating, loss of pleasure in life), are feeling physiological distress, or have suicidal thoughts, it's crucial to seek the help of a psychologist, psychiatrist, psychiatric nurse practitioner, or counselor. Treatment is available and effective and should be sought sooner rather than later. You truly need not suffer alone.

Being able to be honest about these longings, regrets, and sadnesses is crucial to personal and communal healing.

FAMILY CONFLICTS

Given the politicization of all things related to the pandemic as well as the heightened stress for those directly impacted by COVID-19, it is not uncommon for survivors' stories to include family conflict. Unlike illness or death wherein a family can rally and be present as things progress, family members of COVID-19 patients often felt completely left out. In some families, one primary information and treatment point person managed care from a distance and in rushed and stressful times, which prevented them from being able to keep other family members fully informed. Making matters worse, a massive proportion of those taken by COVID-19 were, in no way, ill or close to death prior to their diagnosis. The rapid decline of these patients as well as their lack of access to support people meant that many details that might have typically been addressed prior to a death were never put on the collective family table. While it's common for family members and friends to grieve a shared loss in ways that are particular to each person and their relationship with the deceased, the heightened stress, fear, anger, and shock inherent in COVID-19 deaths provided a complexity that increased the chances of conflict. This leaves many survivors estranged from family members and bereft of the support that they might normally offer. Making sense of traumatic loss is an important part of the process of mourning. Very often, we do this by conferring and consulting with others close to the person who has died. We want to weave together a narrative that makes sense and that can help us work all the way through our grief. When we are cut off from family members or significant others, our grieving process is robbed of this opportunity and is particularly complex.

COMPLICATED GRIEF AND TRAUMA

We addressed the concepts of complicated grief and trauma in chapter 2, recognizing that most people who have lived through this time will

---

HANDLING FAMILY CONFLICT

- **Give yourself time and space to grieve the extra wrinkles in your loss.** It may be helpful to connect with other survivors who have similar experiences or to process with someone who won't minimize or encourage you to give up your own needs to "make peace" too soon or at too great a personal cost.
- **Actively work to build new support networks.** Chosen family can be important to engage as you deal with separations, internal conflict, or loss with your biological family. Faith-, volunteer-, and hobby-related communities can offer vital sources of support.
- **Getting help with navigating the conflict.** For those survivors experiencing this kind of fallout, exploring your reality with a therapist would be a good place to start. Over time, if you feel as though you have interest and the emotional stability to address the conflicts within your family, this professional will be able to help.

---

experience some forms of one or the other. For survivors, however, the likelihood of facing both is heightened.

The experience of surviving trauma can be understood in this metaphor originally set forth by Harvard professor and author John Ratey.[3] We touched on this in chapter 1, but it's so particularly relevant for survivors that it's repeated here. Life before the trauma is akin to walking through a field, completely enthralled by the experience. We notice the sky and feel the breeze. We smell the fresh grass and don't think a thing of noticing a stick on the ground next to where we are standing. Within a moment, our ankle is throbbing and we sense adrenaline coursing through our bodies. We look down to notice that the stick is not a stick but is, instead, a snake. Later on, any time we notice sticks in our pathway, our bodies are likely to respond as though the stick is a snake. We've encoded the trauma in our very cells, which propel our bodies into fight, flight, freeze, or faint mode when we sense threat. Threat, in this scenario, is represented by sticks.

It's easy to imagine how COVID-19 survivors live in a state of constant fight, flight, freeze, or faint. They see someone without a mask and it's like a stick that their body responds to as if it were a snake. They hear or read a news story on the virus and respond as though they've encountered a snake. Every reminder of the virus calls their body and mind into a state of traumatic response. When we add the complexity and many types of loss inherent in every COVID-related death, it's easy to see the depth of suffering every survivor is likely to experience in all kinds of situations.

Trauma doesn't stay in the past, but, instead, lives with us in each moment. Untreated or tended to it can easily usher in panic symptoms (racing heart, feelings of tightness in the chest, difficulty breathing) and/ or Post Traumatic Stress Disorder (flashbacks, nightmares, extreme anxiety, plus symptoms of panic for many). All of these weighty conditions have treatments that are effective. It is important to seek out the help we need, as early as we can, to avoid finding ourselves stuck in these spaces.

DEALING WITH COMPLICATED GRIEF AND TRAUMA

- **Consider the helpfulness of getting answers to unknowns.** Some survivors have found it helpful to request and read the medical records pertaining to their loved one's death in order to fill in blanks from the time of loss. Sometimes, information helps.
- **Develop a mindfulness practice.** Such a practice will help you place a pause between your trauma response and the way you work with it. Basically, learning to pause and connect with yourself will help you discern sticks from snakes.
- **Find grief and trauma specialists to help.** Hospitals, hospice organizations, and places of worship are good places to look for grief groups. The resource chapter in this book will also help.

ANGER

There are many big emotions that survivors are left to contend with, and one of the most common is anger. There's just so much to be angry about. The initial underwhelming response to COVID-19 by national leadership, the hesitance of many to still acknowledge that it is real, and finding that people we have respected are unwilling to follow safety protocols happens every day. It's a lot.

Anger is a normal, often healthy, emotion that can catapult us to important action. It alerts us that there are things to be worked through and may fuel physical actions and efforts that can help make the world better.

DEALING WITH ANGER

- **Make a plan to work through anger and commit to employing it.** Anger must be named and expressed in order to be worked through. The American Psychological Association lists the following actions as strategies for working through anger:*
  - ○ **Relaxation.** Learning some breathing techniques can go a long way to developing a pause between our feelings of anger and taking action on them. Visualizing calm, safe spaces and practicing taking ourselves to them in our minds can also be a helpful tool.
  - ○ **Cognitive restructuring and changing our self-talk.** When we are angry, we tend to think in sentences like, "This is the worst thing ever. This is the most egregious thing that has ever happened." While there is likely some truth in what we think, we also can practice shifting our attention to other truths and offer them to ourselves for help. For instance, "This feels impossible to get through *and* I've gotten through other really hard things. I can be here for myself" helps us feel like we are comforting ourselves rather than working ourselves up. Finding mantras that offer affirmation to understandable anger and that also re-center the potential for calm can also help. For example: "My anger is righteous. Horrible things have, and are, happening. Even still, I can find the kind of comfort that will help me keep going. I can work for justice in large and small ways."
  - ○ **Change environments and avoid hot spots.** If reading or watching the news makes our blood boil, we should try to avoid it . . . at least some of the time. Similarly, if clutter in a certain space of our house makes us spiral, steer clear of that space when on edge.
  - ○ **Use humor.** Humor and laughter are powerful elixirs in treating anger. We are not referring here to using humor to deny or repress anger but, rather, to offer it an equally powerful antidote. Watch a funny movie or read a comedic book or spend time with friends who laugh.

---

* "Controlling Anger—Before It Controls You," American Psychological Association, accessed March 8, 2021, https://www.apa.org/topics/anger/control.

Channeling anger into activism and letting it generate passion for justice and equity is what fuels much of the change in our world. When, however, it becomes excessive it can be disruptive to our ability to calm and comfort ourselves, create stress in our relationships, and negatively impact our physical health. Unchecked, it can feel like it's taking over our lives.

The frequency and intensity with which anger raises its head for survivors can be experienced as highly overwhelming. At times it is so strong that survivors feel as though they can't contain it or turn away from it. Without safe ways of working it out, it becomes toxic to their physical and mental health. Even when they are reassured that anger is an understandable and appropriate reaction to their experience, they often feel perplexed by what to do with it all.

RESPONDING TO PASSING OFFERS FOR SUPPORT

When people we know are facing hard times, it's very common to toss out the phrase "Let me know if there's anything I can do." For COVID-19 survivors in particular, this offer feels complex and triggering. First, there are the somewhat sarcastic responses that flood survivors' minds. "Well, you could start by covering your nose with your mask" or "You might have reconsidered that trip to Hawaii, or, at the very minimum, posting all the photos of you on it with no mask in sight." After this immediate response, survivors are often left not being sure about the sincerity of the offers.

The needs for all people are great as we re-emerge from quarantine. The needs for survivors, however, are great, specific, and pervasive. It's important to identify what needs might be addressed by asking for help from those who have already offered or who we know would want to participate in our healing.

LOSS OF RITUALS AROUND SICKNESS AND DEATH

David Kessler says, "Grief is what's going on inside of us, while mourning is what we do on the outside."[4] This being the case, everyone who has lost a loved one in this time has had their mourning severely inter-

RESPONDING TO OFFERS OF SUPPORT

- **Perfect and practice some responses**. Try actually asking for what you need when a safe and reliable person offers even passive support. In order to be ready to do this, have a varied list of needs in mind. Suggestions include:
  - "Thank you so much for your offer of help. What you could do is to share my story/my loved one's story and encourage people to take time to genuinely and intentionally honor the lives lost to COVID-19."
  - "It's meaningful to have you offer. Might you be willing to drop a card in the mail for my dad? He's really been struggling since Mom died. This would be a big support to both of us."
  - "I'm so grateful for your offer. There actually are some things that I could use some help with." (Ideas would be: a check-in call every once in a while, a meal, a donation to a COVID-19–related nonprofit, or, if you've lost income due to having the virus, a small monetary gift to your Venmo account.)
- **When appropriate, clearly express your needs**. If you are just now waking up to some needs and have had people passively offer since your illness or loss of a loved one, it's perfectly acceptable to circle back around with folks whose offers you declined. In one of my weekly COVID-19 support groups, a person who had suffered with the illness expressed feeling overwhelmed when people previously had made opaque offers of help. Now she could really use the help but hadn't been able to figure out how to say yes to their offers. If this is the case for you, you might say something like, "As things are sort of settling, I'm finding myself grateful for your offer of help. I couldn't really articulate anything then, but now I'm feeling pretty isolated and aware of the fact that I could use connection and comfort in the form of calls/texts, simple expressions of care (a card or cheap bouquet), or small contributions to my PayPal since I've been out of work." When this group member employed this, people were grateful to be asked and excitedly came through for her. It was a beautiful thing to behold.

rupted by the safety requirements in place during the pandemic. It has not been possible to gather with others in person and to receive the kind of physical support that hugs and handshakes or an empathic gaze offer.

For many who lost loved ones at the beginning of the pandemic, there was no physical act of closure in getting to be present for their loved one's last moments or to sit with their body after they had passed.

Rituals are important in that they mark time and offer psychological help in noting beginnings and endings. They also serve as place holders wherein we slow down to integrate important happenings. Our bodies and brains hold the memories of these important times largely around the ways in which we marked them, and we can look back upon them for comfort or help in processing our grief.

In much of the world, when a person is sick, neighbors, friends, and families have rituals that give them behavioral ways of helping. People make chicken noodle soup or bring the patient flowers or magazines. When someone dies, the community comes forth in similar ways, offering tangible support in the way of food, flowers, money, contributions to causes, and physical presence to those grieving. Bereft of these opportunities due to safety protocols around social distancing, losses can feel "unmarked" and/or under-recognized. To the griever it can seem as though the person's life didn't matter to the outside world and, in some cases, can even be experienced in an odd sense of fear that, perhaps, the person never existed at all.

As the world re-enters and people begin to seek a new normal, those whose losses have gone unmarked may feel triggered by the over-

---

MEMORIALIZING OUR LOVED ONES

- **Plan physical memorials when they are safe to hold. Invite others to participate.** If there are ways to memorialize these losses, even months after they occurred, do so. Invite a few close friends to participate in this with you and, if creative, find ways that they can fully participate (everyone lights a candle or offers a word, or the group creates something together in honor of the lost life).
- **Consider an annual way in which to mark your traumatic loss.** Make note of anniversary dates of diagnosis and death and mark them on your calendar. Invite others to participate in a small ceremony and plan for extra self-care. It is normal and healthy to struggle when anniversary dates hit; carve out time to be gentle with yourself around these dates.

excitement of the rest of the world upon re-entry. Their losses may feel diminished and forgotten as the pandemic as it was fades from our collective consciousness. For this reason, it is especially important that we all find ways of marking our loved ones' deaths and acknowledging the significant losses of others regardless of how far after the fact it is.

RITUALS, CELEBRATIONS TO "CATCH UP" ON

It isn't just funerals and memorials that we are behind on or were deprived of in this time. Graduations, weddings, retirements, proms . . . entire years of our lives have gone without any of the celebrations or rituals we might normally engage in. Survivors may not begin to wake up to this reality for some time. In getting through all the holidays and moments of the first year without a loved one, it's likely that many celebrations were painful or entirely missed due to the overwhelm of their grief. This may not seem important to them, however, it may be a source of pain for those close to them who either missed their full presence in the past year or who feel somewhat "untended to" due to the survivor's immersion in their own grief.

It can feel overwhelming, in these cases, to think about how to repair or address the feelings of those in their relational sphere. It will also be important for survivors to watch for feelings of resentment toward those in their close circles as they celebrate moving on from the pandemic. These complex emotions make sense and require attention. The complexity of every person's feelings in these relational circles may make communication difficult, but it's worth it to persist in efforts to keep trying.

Starting with a commitment to using "I" statements and a reflective listing technique can be a way of setting the stage for everyone to be safe in these kinds of conversations. This means that all participants use their own feelings and experiences to center the conversation rather than pointing out what the others have done "wrong." This would mean that a child might say to a grieving parent, "I felt really sad when we didn't get to celebrate my birthday because grandma died." The parent might then respond with, "I understand how disappointing that must have been. I really get it. I was hurting a lot at that time. Can we find some ways of marking your birthday today?" This doesn't mean that the grieving party takes on all of the feelings of the person who feels slighted. It simply makes space for each person to make their feelings known.

Finally, it may be that survivors have been, or have felt, failed by their communities. They may harbor anger or resentment resulting from feeling as though important people in their lives did not show up for them in the ways that they might have before the pandemic. The same kind of reflective and responsibility-taking suggested above can help guide these conversations. For example, one might say, "I need to talk with you about my grief process. I know that quarantine impacted the ability for people to be with me in the same way they might have been otherwise. Even still, I have some feelings of being forgotten. I think I might need some active support to feel seen. Are you able to offer this?" While this conversation starter might take immense courage to speak, it would go a long way in helping the healing begin on even footing.

CATCHING UP AFTER A TRAUMATIC YEAR

- **When emotionally ready, start conversations**. When you feel ready, start conversations about any events or experiences you've missed that you'd like to catch up on. Be careful when asking if there are ways you haven't shown up for others. There's a fine line between being open to honesty and shoveling guilt and shame upon yourself. It's crucial to remember that you can hold understanding and grace toward yourself without taking more responsibility than is yours.
- **Consider hosting a small "catching-up" ritual or experience with yourself or others**. Set aside some time to look over your calendar and notice the things you've missed out on since your diagnosis or the diagnosis of a loved one. Write short texts or emails to those whose big moments you've missed. Use these to honor the people, not to be overly apologetic about the space and time you've needed this year.

SERIOUSLY "UP" YOUR SELF-CARE GAME

The profound weight of the pandemic on survivors cannot be over-estimated. Neither can the emotional complexities they will face in witnessing the world re-opening, diminishing the global attention to COVID-19. It would be like someone losing their personal freedoms and loved ones to cancer for a year and then the whole world saying, "Yay! Cancer's over. Let's put it behind us and never speak of it again."

A strong and realistic self-care plan can be of great help to survivors. This shouldn't mean manicures and happy hours alone. It should, instead, be a well-thought-out plan for maintaining a deep commitment to mental and physical well-being and a consistent tending to one's relational needs. Self-care plans do nothing for us if we don't create them with reality in mind or make commitments to practice them consistently. The following ideas can help survivors (and everyone) identify ways of caring for themselves into the future.

TENDING TO SELF-CARE

- **Make a self-care list**. Make a list of twenty diverse actions that are deeply self-caring to you. Some should take mere moments to enact (Take five diaphragmatic breaths.) and others should require more time and planning (Go to the beach for a day.). Some should be free and easily accessible (Watch a relaxing show. Go for a run/walk.) and others might require a small investment (Take a yoga class. Take a class for a hobby I'd like to pursue.). Include at least one thing that you'll really need to dream about, plan for, and work toward. Post this list where you can see it. When needed, look at the list and choose something without letting yourself labor to find the "right" or "perfect" thing.
- **Enlist self-care partners who are healthy and will maintain your boundaries**. Share your list, or a few items from it, with a trusted friend or colleague who can encourage you to take a break when you need one.
- **Plan ahead for times you know extra self-care measures will help**. Looking ahead at your schedule, make notes on your calendar about when you'll need more self-care. Plan the time for it.
- **Put monthly reviews or bi-monthly reviews on your schedule**. Put a review with yourself on your calendar just like you would with an employee or employer. Don't let yourself cancel or overlook this time to look back over what is and isn't working in your self-care plan.

4

RE-ENTERING THE WORKPLACE

Search "re-entering the workplace post COVID-19" online and thousands of clickable links will appear. Most of them list actions to be taken by vocational settings in order to assure the physical safety of their clients and employees. One of the best that I've come across lists the following actions:

1. Make a set of "no personal contact" rules.
2. Encourage no "item sharing" when possible.
3. Reorganize your floor plan.
4. Get rid of common gathering areas.
5. Adjust break room rules.
6. Create prominent hand-washing stations.
7. Post communal-equipment cleaning rules.
8. Create appropriate face mask rules.
9. Limit the number of people in a closed room.
10. Break the 9 to 5 by adjusting work hours.[1]

It's clear to see that work settings may not be "back to normal" any time soon even though we may be back to working in-person right away. This return to an evolving unknown is likely to be unsettling for both the employer and employee, requiring flexibility and a keen eye on balancing the needs of the employees with the needs of the population that the business serves.

Healthy organizations share certain traits that range from effective teamwork to high morale as well as integrity and a commitment to a willingness to see mistakes as opportunities for learning. When a work community is healthy and thriving, it supports the economic needs of its employees while also helping them to find personal satisfaction and meaning. Social connections and support are also benefits of corporeal work spaces. For an organization to stay healthy and relevant, its workforce must feel valued and cared for, and when this happens, the organization benefits. This mutual give-and-take is crucial for both parties (the individual employee and the business) to remain content and high functioning.

This chapter addresses the psychological and mental health impact that business communities, as well as the individuals managing and working within them, will be facing as the world re-opens. Organizations will face unique challenges and will be benefitted by active forethought and intentionality as they reimagine post-pandemic work spaces.

RECOGNIZE THE IMPORTANCE OF VOCATION AND WORKPLACE SETTINGS TO MENTAL HEALTH

Long before the pandemic required a majority of workers to migrate their offices to their homes, researchers were finding that "fast-paced work, continuous demand to learn and use newer technologies,[2,3,4] and reduced people interaction [were] all causing significant stress on employees, placing higher demands on employees' well-being, and in turn, on the health and efficacy of organizations."[5] It's fair to say that the race to find new ways of doing business when quarantine orders rolled out required everyone to learn and use new technologies and reduced people interaction in profound ways. All workers and businesses will feel a cost for this.

Research suggests a strong and significant negative correlation between job satisfaction and psychological distress as evidenced by insomnia, depression, anxiety, and hostility. This means that individuals reporting dissatisfaction at work experience higher levels of psychological symptoms than those who claim higher satisfaction at work.[6] This is important to pay attention to at individual and corporate levels. When we feel satisfied at work, we gain a sense of accomplishment and meaning. These are part of our "pay," so to speak. When we don't feel satisfied, we pay a personal price by experiencing distress. Corporately, high job satisfaction in individual workers can be an important predictor of the overall health and productivity of organizations. When workers are satisfied the business thrives.

The pandemic has had a severe impact on people's relationship to the work they do, how they do it, and how they interact with their peers. For those who thrive in team settings, the loss of the dynamic energy of a people-filled work space has been profound, leading many to dislike work that they previously enjoyed. For those who prefer autonomous work within the context of others, the need to schedule meetings and hold them via video chat, as opposed to consulting in organic and spontaneous ways, may have taken an emotional toll.

Further, the "always available" expectation that comes with having our work space in our living space has had a major impact on our collective and global psyche. We aren't sure where the boundaries can

---

TENDING TO THE RELATIONSHIP BETWEEN JOB SATISFACTION AND MENTAL HEALTH

- **Do an examen about how your work gives you life and takes it away**. On the front of a piece of paper write "Work Tasks That Are Satisfying/Life or Energy Giving." On the back, write "Work Tasks That Cause Distress/Take Undue Life or Energy from Me." Set a timer for ten minutes and brainstorm each category, writing down each task in its rightful list, omitting nothing. Look over the list and write a feeling or two that you associate with each of these tasks. Acknowledge the painful feelings associated with the tasks that lead to dissatisfaction. Identify any small tweaks that might be possible to either reframe your approach to the tasks or ways that you might work with your manager or team to get assistance or swap tasks with others.
- **Determine the way in which your values match or do not match with those of your employer**. Using a values assessment you find online, identify your top four to five values and then compare them to the values in your employer's mission statement. Notice the way in which they mesh or clash. Determine any ways in which you might be able to align your personal work within the corporate setting with your personal values. If this is impossible, begin to dream about work that would offer a better fit and begin taking small steps to make it possible for you to seek this out.

and should be constructed anymore. We know that people are home, we know that they are carrying their devices, so we have come to passively expect that work will be done immediately, regardless of the time of day or night. In holding ourselves and our coworkers to this, we pile stress upon stress about our response times and keeping up. This is only one of the profound shifts in our ideas about work that the pandemic has brought about.

Owning that our work feeds our sense of well-being as much as it does our bank account is crucial for understanding the profound way in which our return to an entirely altered work reality might impact us. Keeping an eye focused on our level of psychological distress in relationship to our work will help us address issues that might lead to psychological symptoms early enough to prevent them from taking root.

ACKNOWLEDGING WHAT IS LOST AND THE GRIEF THAT RESULTS

If we reference back to the beginning of this chapter and the list of changes businesses will need to make upon re-opening, we'll notice that there will be profound losses in the kinds of experiences we might expect to have at work. In the earliest stages of re-entry, we'll return to offices bereft of gathering places and break rooms. Our desks will be farther apart and large team meetings will continue to take place online even when we're all in the same physical location.

Perhaps most importantly, any ease and familiarity we enjoyed in our workplaces before the pandemic will likely be impacted by the need to maintain distance and retain a certain vigilance about shared spaces. This, in and of itself, comes with a price. Days that were previously peppered with lighthearted interactions, casual social interaction, or meaningful banter will now be infused with a carefulness that could easily disrupt calm.

It's important to grieve the losses inherent in all of this. Grief isn't an emotion that simply goes away. Unacknowledged and unexpressed, it can have a serious impact on our mental health. Grief is rarely an easy feeling to work through, especially for those people who experience a greater distance from their felt emotions. While grief has shared characteristics (sadness, anger, denial, acceptance), it doesn't take a linear course. Grief pops up when least expected and is difficult to completely avoid as it makes itself known via the body (health concerns, pain, etc.) if it isn't addressed. Losses also knit themselves together, creating a compound source of grief made up of all unresolved/unaddressed sources of loss.

If we can, both individually and corporately, acknowledge the feelings associated with that which is different and/or missing in our work spaces as we return, and provide space for supportive grieving, we are much more likely to withstand the return to our new normal without experiencing psychological distress.

To do this, it's important to be forthright and honest about the personal disappointments that will result from instituting changes that require a certain level of vigilance and isolation in the workplace. Instead of trying to "rally the troops" and "get them excited about the new workplace," managers and executives will be better served by facilitating openhearted conversations wherein they acknowledge the real losses that they are asking their employees to endure. This need not be a dramatic process, but rather can include an acknowledgment followed by expressions of empathy and appreciation.

It's important to recognize that a proportion of the workforce will face feelings of grief in being asked to return to physical workplaces. Many people have enjoyed working from home and have made significant investments in making this work. Whether these are monetary (investing in office furniture, etc.), relational (working from home has significantly increased the time spent with partners and/or family given the loss of commute time, etc.), or primarily personal (work from home allows fewer distractions than a shared work space, etc.), the loss of these secondary payoffs is real and deserves to be grieved.

In order for all people facing the very real losses of parts of their work that they enjoyed to stay healthy, taking active steps to grieve will be crucial. It may feel odd to give emotional energy to working through feelings related to losses at work but it's important to recognize that doing so will make for a much healthier transition and happier employees.

ACKNOWLEDGING GRIEF

- **Recognize the need to grieve**. Every person's grief process will be different, but everyone will benefit by naming the multitude of losses and normalizing the varying emotions associated with them.
- **Get honest about the losses you've faced/are facing**. Take some time to make a list of the losses that you experienced when COVID-19 required the world of work to change. Now, make a new list of the losses that you anticipate you'll experience as the world re-opens. Find ways of expressing these and working them through.
- **Provide resources to your employees**. If you are an employer or team manager, offer grief education and validate the losses employees face. Hire a psychologist or grief specialist to do a short (fifteen-minute) talk on grief and then facilitate a conversation with employees. Send a personal note acknowledging all the changes and flexibility you've required of your workers and offer some resources such as EAP visits or an hour of PTO for an employee to go to therapy. Tending to employee health will be a huge investment in your business's ability to manage another huge set of changes. Remove stigma around your people doing what they need to in order to maintain their physical and mental health.

ACKNOWLEDGING THE HABITS THAT WORK FROM HOME HAS CREATED AND REINFORCED

It's hard to fathom the number of changes to a "normal" workday that the pandemic has forced, with myriad small rituals and routines tossed aside the instant the world shut down and we began working from home. Where we would normally have risen and dressed, engaged in rituals around making coffee or tea, and then stepped outside to walk, pedal, or drive to work or school, we suddenly became able to roll out of bed and into a meeting. We only needed to be half dressed and, when breaks came, we toggled from our work screens to the online spaces that entertained or informed us without ever moving from our seats. When the end of the workday hit, we did the same, never really marking time with the kind of meaningful rituals that would significantly tip our bodies off to the needs for rest or transition. As a result, we may have worked too hard or too long or had a harder time focusing on work than we ever had when it was done in a completely set-aside space. For some, their performance dropped altogether. For others, their work knew no bounds.

For many people this meant that geographic spaces once demarcated specifically for rest and family time (homes) were no longer spaces dedicated to and saved for nonwork-related activities. Not only were home interiors changed, but the introduction of desks and monitors and all other manner of office equipment blurred the lines between work and home in a way that will be hard to come back from. Where we might have logged out of our work email when we left the office, the constant reminder of having our work computers in our living room or bedroom will make it harder to completely separate from work like we used to. Because others knew that we weren't out and about, engaging in our pre-pandemic activities, we felt compelled to answer the phone or emails after hours in ways we might not have before sheltering in place. This was all made more complicated by the fact that we know that our coworkers are feeling, and succumbing to, the same pressure, and we feel we need to measure up.

We must also acknowledge that moving work to home offered new opportunities for multitasking that may have added to our sense of overwhelm. Simply because we could do laundry between meetings or fix dinner or work out while on calls, we did, and we embraced this ability with abandon. This blurring of lines between work and home life has taken a toll and will continue to do so. Employees will be entering the workplace with varying levels of insight about this. Some will have worked through the way in which they will manage the shift back to commutes and shared space but others will not and may not be able to keep up the pace expected of them when they were working at home. This creates an elevated risk of rises in work-related stress, which could, eventually, harm workers and the larger work space.

ADDRESSING OUR WORK/HOME HABITS

- **Renegotiate the work/life issue**. If you are a manager or team lead, start honest conversations within the entire organization about how to reset norms about work/life balance. Find ways to honor employees' need to set firmer boundaries around away-from-work-hours requests and have conversations that get everyone on the same page.
- **List your pre-pandemic daily rituals**. Identify the small and large rituals that marked your days pre-pandemic. What were your morning routines like? What was it like to "clock in" to work? What signaled the end of your workday? How did you separate yourself from work demands outside of work hours?
- **List the rituals that emerged during work-from-home orders**. How do your morning routines signal a transition of personal time to work time? In what ways do you (or do you not) take breaks from work during the day? How do you mark the end of the workday and re-entry into personal time?
- **Explore ideas for protecting personal and work time from each other**. How often do you find yourself checking your personal emails/socials/and so on during work hours and vice versa? Identify the costs, if any, of not checking on work in off hours.
- **Distinguish boundaries between personal time and work time**. Brainstorm some routines to put in place to mark the beginning and end of work and private time. For example, turning off work message indicators when arriving home. Find ways of offering yourself psychological markers of time intended for work or for personal needs.
- **Commit to ten-minute breaks from devices and work**. Set two ten-minute times per day to take full breaks from devices and two short times (one in the morning and one in the afternoon) that you can check personal email and socials during the workday. Set alarms for these actions and commit to honoring them for a full week.
- **Set the norm of not checking work emails outside of work hours**. Determine what it would cost you to not check work email once work hours are completed. If the cost is one you can afford, make this commitment and uphold it. Deleting work email from smart phones can help with this as can using only one computer for work email and powering it all the way off each evening.

ADJUST RELATIONAL EXPECTATIONS GIVEN NEW STANDARDS/REQUIREMENTS

Very frequently, work settings offer the fringe benefit of social opportunities. Whether these occur around the (proverbial or actual) watercooler or in meeting rooms or they transfer out to offerings such as work league sports teams or volunteer shifts, at least some of our coworkers are likely also our friends. The opportunity to re-gather in person at work, then, is likely a cause for celebration for many. For this group of people, adjusting to the new realities of the workplace may be difficult. With break rooms off-limits and distancing measures in place, it may feel awkward to be back in physical proximity of others without being able to resume pre-pandemic types of connectivity.

As employers reconfigure work spaces, they would do well to keep this in mind. In climates where outdoor tables and chairs could offer appropriately distanced and ventilated space for small groups of coworkers to congregate, they would be a helpful addition. In settings where break rooms and kitchens previously offered up shared snacks, creating individually wrapped offerings might maintain the feeling of community while also honoring safety protocols. Anything that approximates an embodied sense of community will help employees feel connected to their work and each other.

RESTARTING EMBODIED WORK RELATIONSHIPS

- **Create safe places for small groups to meet.** Offer outdoor tables and chairs for small sets of employees to socialize.
- **Establish a five-minute daily "bonding" experience for employees.** Institute a short daily video/in-person conference for all employees. Offer a bonding experience to everyone. Play a song for everyone to have "stuck" in their heads all day. Have someone tell a joke or offer a pep talk. The goal is for everyone to feel "gathered" as part of a community at least once a day.
- **Build/make a "give or take" wall or encourage notes of gratitude and support.** Designate a wall where people can leave words of encouragement, humor, or support written on sticky notes. People can offer whatever words they have to encourage or brighten a day, and others can take those back to their work spaces for encouragement.

OFFER DIRECT SUPPORT

The fact is that, no matter how much preparation employers and employees do as they re-enter more widely shared spaces, everyone is likely to experience distress and missteps. Someone is going to be so lost in glee about seeing their former colleagues that they go in for a handshake or hug without gaining consent. Another will joke about "this nightmare finally being over" to a coworker who lost a parent to the virus. Employers will fail to reinforce safety guidelines and employees will feel stress about how to deal with this without support and so on.

It will be wise for each of us to have a plan about how we will weather the inevitable foibles and failures of re-entering the workplace. For those in positions of leadership, the establishment of clearly communicated guidelines by which people can register concerns will be necessary for all parties to feel safe in navigating the murky waters of re-establishing norms. Nonreactive yet appropriately attentive responses to expressions of discomfort or reports of missteps will ensure that all voices are heard. These will take effort to create and establish.

Individual employees returning to workplaces would also do well to do a bit of planning. Determining, ahead of time, what one is and is not comfortable with, how and where to register concerns, and who they can go to if they feel at risk will help employees be ready to step back into the workplace with relative confidence.

PLANNING FOR AND OFFERING SUPPORT IN THE RETURN TO PHYSICAL WORK SPACES

- **For employers.** Consider appointing a person or set of people to field employee concerns about the actual space and compliance with protocol in a trauma-informed manner or host support circles/groups at the end of each of the first four weeks of returning to work. Hire a therapist or Critical Incident Stress Debrief professional to host these groups. Remembering that a workplace can only be as effective as the collective health of the employees, privilege their care and concerns in this time. The payoff will be great.
- **For employees.** Make lists of what you've gained and lost while working from home as well as what you're dreading and looking forward to. Pay attention to the feelings associated with each and begin to set realistic expectations about how you'll work through the unknowns.

5

HELPING CHILDREN RESTART

A ll young people, and the grown people attached to them, are exhausted and overwhelmed. Parents and teachers have done their best to help them through the constantly changing, yet somehow always monotonous and impossible, seasons of the pandemic with varying levels of success. Parents beat themselves up for their perceived failures, worry about the long-term impact of quarantine on their kids, and feel on the edge of losing it. Children are anxious and depressed, sick of video calls, and, generally, experiencing a weird mix of exhaustion and agitation. Everyone's tired of each other.

Families who had babies right before or during quarantine have done these early months all alone, without the communities that would normally step in to help. Grandparents and extended family may have been angry for being "shut out" due to safety precautions, and everyone is irritated. These families will be teaching their children an entirely new way of being in the world . . . with others.

The back and forth of online and in-person school, access to reliable technology and internet, and the fact that students have had more screen time than would ever have been considered "sane" will make for a difficult transition for school-aged children and their parents. Separation anxiety or impulse control in re-entering may be issues, as will the reality that many students have fallen behind academically and/or socially in the pandemic year.

Parents with adolescents and young adults have likely hit walls over and over in this time. No one was meant to spend their adolescence cooped up. These months of re-entering wisely will be challenging.

All families are about to go on the wild ride of continuing to tend to the impact of all that they've been through during quarantine while also helping their children and themselves move forward into the unknown. This will be challenging!

While the needs will be different for each family, there are some similarities that all will face. These are listed here with ideas for how to practically help young people re-enter a wider embodied world. Each section includes tips and tools as well as conversation starters for families with children of all ages.

PARENTS ARE AT THE END OF THEIR ROPES

In her brilliant *New York Times* opinion piece, author Jennifer Senior writes about the way in which COVID-19 has brought parents—mothers in particular—to a place of intense emotional dysregulation. She writes, "We're in the midst of a global crisis that seems almost perfectly engineered to make us meaner. We're cooped up. We're isolated. And as I wrote late last spring, we can't find flow—not while working, caregiving, cooking, cleaning or even watching reruns of bad TV—because the demands of the kids, the house, the job (if we're fortunate enough to still have one) collide with one another, subdividing our days into staccato pulses of two-minute activities before we switch to something else. It's all disruption all the time. . . . It's perverse. The pandemic has made a thing that was already a source of shame for many of us all the more acute."[1] It's fair to say that parents will be doing the hard work of re-entry from places of pure exhaustion.

Mothers appear to have taken an especially hard hit, feeling as though they weren't capable of living up to what was required of them in this time. In December of 2020, the *New York Times* opened a hotline for mothers to call to vent their COVID-related grief. Hundreds of mothers called in, yelling, screaming, and swearing as a way of trying to release their feelings. Journalists Jessica Grose, Farah Miller, and Jessica Bennett began a project wherein they followed three mothers who wrote extensive journal entries and stayed closely in touch from September through December of 2020. In February of 2021 the journalists (all mothers themselves) told these women's stories, which were thought to be representative of all mothers, in seven "chapters": Chaos, Resignation, Drowning, Exhaustion, Resentment, Perseverance, and Hope.[2] These chapter titles tell the whole story of what it's been like to parent in this time.

To be honest, parenting has never been easy, and the way in which we parent often comes from a strong reaction to the way in which we, ourselves, were parented. If we grew up in a culture of detachment and criticism, we are likely to raise our children from the same stance. Too often, we either fall into certain patterns because they are familiar to us or we live in reactive denial of our personal pasts and make parenting choices reactively. Research shows that parents who have taken at least one step to work through the way in which they were raised (such as go to therapy, participate in a parenting group, read books and discuss them) raise more securely attached children. This is important because attachment will help children weather difficult times like the one we are in. Daniel Siegel says, "When it comes to how our children will be attached to us, having difficult experiences early in life is less important than whether we've found a way to make sense of how those experiences have affected us. Making sense is a source of strength and resilience."[3] Since it's been hard for us, as grown people, to make sense of the realities this year, it's likely that our children are doubly impacted. They see us floundering to compensate in and of ourselves as well as feeling our limits in terms of helping them stay regulated.

Making matters worse, job loss and economic hardship have faced far too many families in this difficult year. The stress of unemployment or of overstretched budgets is especially toxic and can leave parents feeling incapable at entirely new levels. Understanding the weight that this has is important so that these parents can begin to give themselves a break for all that they are carrying. Redirecting their attention to the incredible gifts that they do give their children can help but is especially difficult when life feels so overly full of stressors. It's clear that parents are in need of a significant break right at the exact time when their children will need a lot of help navigating the big changes ahead. Here are some ideas for helping parents catch their breath.

THINGS TO TRY: PARENTING WELL WHEN EXHAUSTED

- **Direct your attention to parenting strengths**. Make an exhaustive list of ways in which you succeed as a parent. Include things like, "I prepare meals for my children" (avoid judgment-based words in the sentence such as "healthy" or "balanced") or "I tend to my child's hygiene." Look at the list regularly to help yourself hold the entire continuum of your parenting. When tempted to obsess about parenting failures, acknowledge mistakes, repair, and then move forward, focusing on the ways in which you effectively parent your child.
- **Focus on a moment of "flow."** Identify a time when you and your child were in a great "zone," where you were connected and things were flowing smoothly between you. When you begin to beat yourself up, give yourself one full minute to immerse yourself in this memory.
- **Give yourself a time-out**. Begin to notice the early stages of losing your cool and practice inserting a pause between that experience and acting on it. Use language like, "I need to leave this situation for a minute to get myself calmed down."
- **Be ready and willing to say "I'm sorry."** When you do make a mess emotionally, boldly apologize. Be as straightforward and heartfelt as possible. Resist saying things like, "I'm sorry you felt hurt" and say, instead, "I'm sorry that I hurt your feelings. What can I do to repair this?"
- **Do at least one thing to address parenting habits that are reactions to the way in which you were parented**. If you are finding yourself stuck in parenting habits that are hurting your ability to attach meaningfully to your child, do one thing to address that. Go to therapy, read a really good parenting book, or find a parenting class or support group. Your entire family will be rewarded.
- **Offer yourself grace and reward your hard work**. Find ways of rewarding yourself for keeping going. Take a nap or go for a walk alone. Eat your favorite foods for a day rather than preparing your children's favorites. This is hard work and small rewards help us keep doing it.

COVID-19 HAS BEEN STRESSFUL AND SCARY FOR KIDS

The systemic ills of the world and vulnerabilities of life have been laid bare during this time, and this has not been lost on children. Quarantine has brought these into dramatic relief for the youngest among us. Children facing food insecurity haven't been able to receive meals at school or partake from high school and college food pantries. Opportunity and resource gaps that existed for children growing up in economically disadvantaged homes, for people of color, and non-English-speaking families have had increasingly dire consequences. And children whose parents have a greater difficulty with emotional regulation have been sheltering in unsafe quarters, unable to check in with teachers and counselors who have served as adult mentors.

Even for families privileged not to fall into the above categories, life has been stressful and scary. Children have had to be careful in ways we never dreamed they'd need to be, and even loved ones have been kept at a distance. Parents have done their best to keep their own anxiety under wraps, but most children have "felt" their parents' distress. In many ways, all innocence has been lost, and the fact that COVID-19–related fear is real and understandable makes it difficult to work with.

We can't, usually, make promises that a person's fear won't come to fruition, even if it's a long shot. We can, however, help them understand that fear is their mind's way of telling them to pay attention, not, necessarily that something is wrong. The best way of helping people deal with fear is helping them develop tools for getting through it, *not* avoiding altogether. When they feel fear, they can look around to see if there is something they need to do. If not, the best course of action is to calm themselves by redirecting their attention or self-soothing. We'll address self-soothing in the following section.

Redirecting attention in response to a COVID-19 fear might sound like, Child: "I'm afraid of getting the virus at school." Parent/Adult: "I understand that. We've had to be super careful. I also know that you know lots of things that you can do to 'talk back' to your fear. What are some things you *can* do to stay safe?" Child: "I can wash my hands and wear a mask." Parent/Adult: "Absolutely! So when you feel that scared feeling popping up, notice it, and then turn your attention to all the things you're doing to protect yourself."

Children are likely re-entering the world with less resilience than they had pre-pandemic. Some may be academically or developmentally behind; others may face crippling social anxiety or phobia-like reactions around safety protocols. All will be overwhelmed. Whether they've had help dealing with the big feelings that have visited them this year or not, they will be doing the work of navigating yet more highly taxing change.

The children's television host Mr. Rogers frequently said, "If something is mentionable, it is manageable." This is so true. Simply helping a child of any age give voice to what they have experienced, naming how it has impacted them emotionally, physically, and intellectually will go a long way toward helping them work through challenges. Witnessing their parents naming and working through their own big feelings will give them a head start. Having those same important grown people help them name the child's feelings comes next. Onboarding a few self-soothing skills and planning how to move forward with them in the tool kit is icing on the cake.

A good conversation can be the best starting point for developing a tool kit for our children, but it can be difficult to know how to have one with our children when the realities they are facing are so huge. Here are some conversation starters to use when addressing children's fears.

For Young Children

"We've gotten through a super-hard time. We've felt a lot of big feelings. Sadness. Frustration. Being angry. We're getting ready to be with more people and in more places. Whenever you notice a big feeling, you can tell me about it. Let's take a couple of big breaths together. (Show inhaling through the nose ["smelling the roses"] and exhaling through the mouth ["blowing out the candles"].) Whenever we're trying something new or you notice a big feeling, you can ask me to do this with you. I'm also going to ask you, now and then, how you're feeling about all the changes in our lives."

For Six- to Twelve-Year-Old Children

"I can imagine that you have a lot of thoughts and feelings about going back to school/having grandma and grandpa over/us being out and about more. Does it feel weird? Good? Scary? (Listen here.) I know that I'm thinking a lot about how different things are and I'm noticing sometimes I feel differently than I normally would. I just want to tell you this because I imagine you might feel a bunch of feelings as things change. I'd love to hear when you do, and I have some ideas of things that can help. Or, we can brainstorm together."

For Adolescents and Young Adults

"You've made it through a completely life-altering time. I am trying to remind myself that it's normal for me to feel a lot of feelings about all the things I've missed because of COVID and to feel weird about everything changing again. What are you feeling? (Notice this is not a yes or no question.) I'm super happy to help you find ways of dealing with all the changes. I'll try to remember to touch base about it, and you can always communicate with me about any help you might want or need . . . in any way you want (text, note, conversation)."

THINGS TO TRY: HELPING CHILDREN WITH FEELINGS OF OVERWHELM AND FEAR

- **Help children work through their fears.** The goal is not to help children avoid fear but to help them develop the tools to move through it. Acknowledge their fears and empower them to confront them with self-soothing in mind. Develop some mantras or identify a soothing object that can help them feel solid as they confront fears.
- **Avoid avoidance.** Offer tools, training, and support for your child as they approach their fears. Do not actively help them avoid them. Instead, believe in your child and help them help themselves.

DEALING WITH OVERWHELMING EMOTIONS AND TEACHING SELF-SOOTHING

When we feel overwhelmed it's easy to become dysregulated. Our heart speeds up, we feel agitated or "jumpy," we can't think clearly, and our emotions are right at the surface. Sometimes the opposite happens and we shut down, we can't breathe well, and our minds go blank. Whichever way we, or our children, go, it's a difficult reality to live through.

One of the ways that we use to get back to a sense of "normal" is by having freedom to leave situations that overwhelm us and get into spaces that help us feel calm. We reach out to someone who will listen or help us laugh. We run errands or shoot hoops or do whatever our calendar reminds us we've committed to and, over time, we find ourselves coming back to center. In this time of shutdown, we haven't gotten to do these things, and we're all less regulated than ever as a result.

This is a scary and uncomfortable state of being for children . . . especially when they have little control. They feel the full weight of their feelings but sometimes don't have language or skills to express them. They feel stimulation will actually calm them so they seek it out nonstop. They're "not tired" (they are *never tired*) when we can clearly see that they can barely keep their eyes open. Their overwhelm causes them to feel vulnerable, which can make it very tricky for them to want to acknowledge. Even if they did, COVID-19 has left them bereft of ways of working through it.

Children have been sheltering with a fixed number of people in only a few spaces, which has severely limited the options they have for getting their feet under them when overwhelmed. They've missed out on a lot of the kinds of physical and social activities that would normally offer them release, and there aren't many non-digital distractions within their reach.

It's important to note that long before the pandemic had us relying on our screens for nearly everything, our devices were shown to create high levels of emotional and physiological dysregulation. This means that too much time with a screen can leave us feeling overstimulated and jittery, agitated and out of touch with our bodies. We've acclimated to such high levels of stimulation, however, that we feel overwhelmed when we try to step away from it. Especially for kids, they're damned if they do and damned if they don't.

For far too long we've substituted stimulation for soothing. We've told ourselves that a couple of hours gaming, or Pinterest, movies, videos, or scrolling through social media at the end of the day "calms us down." In reality, however, all it does is distract us from our dis-ease and reinforce the reality that we don't have what we need to comfort or regulate ourselves on our own. This is especially true for children.

If I want to run a race and my stomach is empty, I could make the choice to guzzle a bottle of water to make it feel full and stop growling. About a half mile into the run, however, my body would begin to realize that I hadn't offered it any actual fuel. Yes, I'd made myself feel full but I would be lacking actual calories to give me energy, and I would go into all manner of dysregulation as I pushed my body to do what it wasn't prepared for.

This is what it's like for children who have substituted in-person encounters in the world (even those as casual as the ones found when walking through the halls, getting a haircut, or attending a school event) with the less energetic encounters we find online (even if it's online school). Our embodied interactions with others, especially important others who we feel are on our team, give us emotional and energetic fuel to get through life. Given this, children are feeling the way it feels when we are pushing hard to do something but don't really have what we need to do it.

It is so important that parents do whatever they can to have immense grace for their children as they re-enter in-person spaces. They're going to act out more, they're going to want more attention, they're going to worry about things that seem ridiculous to us. This is what any of us would do when trying to run a marathon with no training or resources.

Complicating the work of grown people in knowing how to respond to their unrest and overwhelm, children will typically do what they need to do to get an intense reaction from us. They care less, unconsciously, about the *kind* of reaction (positive or negative) than they do about the *intensity* with which we respond to them. They need us to pay attention when they are overwhelmed and they rarely know how to tell us that effectively so, instead, they rage or tantrum or run off to isolate themselves. If they can get a more focused and intense response by acting out or expressing huge worry, their unconscious minds will drive them to do so. In many ways, our full and intense attention feels so grounding that they're willing to risk it being negative. Falsely positive, weak, or unfocused attention doesn't feel like enough when they are overwhelmed and dysregulated . . . which is exactly how they are at this point in the pandemic. Given their own exhaustion, this is going to be hard for parents. Even still, there are some accessible ways through.

Dealing with emotional dysregulation (in ourselves and with our kids) requires that we know what actually soothes us. So, we need to get to work helping our children develop self-soothing plans that are specific to their temperaments, resources, and preferences. For some children this will be learning deep-breathing techniques, and for others it will involve running up and down the stairs until they feel they've released their pent-up feelings. Some children will want to talk about what they are experiencing and others will need to sing, dance, or draw their experiences. For some children, throwing ice cubes at the back fence and watching them explode will work well, while others will want to be held and cuddled. The important thing for parents to do is to notice what really works for their children and facilitate those actions. This may require some direct teaching and/or setting up our homes to offer self-soothing helps.

One way to do this is to employ a color system for family members to communicate how they're feeling. The Zones of Regulation model is a great one, with green meaning, "I feel calm and alert. I'm ready to engage. I feel content."; yellow meaning, "I'm sort of on edge. I am not quite calm but I'm also not super-agitated. If I face stress right now, I'll likely tip into the red zone."; and red meaning, "I feel dysregulated and upset. I can't think straight because of my big emotions." Prior to navigating difficult conversations or experiences, family members can get a self-report of how other members are feeling by ascertaining which color they fall into at that point in time and adjust their approach accordingly.[4]

It may feel odd to ask our families to use such a method or to teach a young adult how to do deep breathing or a meditation. It may feel uncomfortable to bear witness to the intensity of feelings from an adolescent or to invite a conversation about big feelings with a young child. It's critical, however, to push past these feelings of discomfort in order to model and intentionally teach children the art of self-soothing and self-care. Having some general self-soothing and emotional-regulation tools to teach children, however, can be a life changer. Basic practices such as those listed here would be good places to start.

> *Smelling the roses* (inhaling through the nose) and *blowing out the candles* (exhaling through the mouth).

Diaphragmatic breathing (focusing on filling the lower portion of the lungs on every inhale, making the belly expand, and then emptying the same lower portion of the lungs on the exhale, making the belly fall flat. It can help to ask a child to lie on their back with a penny on their belly. On the inhale, they make the penny rise up to the sky and on the exhale, they make it drop down toward the earth).

Creating soothing environments. Offer dedicated spaces in the home where family members can go to calm down. Fill these with things like face-level lighting instead of overhead, weighted blankets, calming music, water bottles, a stuffed animal or non-digital handheld game or puzzle, and some paper and pens.

Begin daily three- to five-minute family time-outs for doing relatively nothing at all. Be bored together or do some breathing or stretching. Boredom is important for emotional regulation and can be a source of deep grounding when practiced regularly.

Having conversations about emotional overwhelm and self-soothing can feel awkward. It's important, however, to have them anyway. The payoff will be huge. Here are some helps for starting those important interactions.

For Young Children

"When you feel upset or scared, what helps you feel better? (List some things you've observed if they can't think of anything, such as letting me read to you, holding a blanket or other comfort item, going outside for fresh air.) As things are changing and we're seeing more people, I hope you'll let me know any time big feelings come up. I want to help you when you feel them."

For Six- to Twelve-Year-Old Children

"We have really gone through a lot together this year. We've seen each other feel a ton of big feelings. Now that we can see more people and do different things, we might end up feeling more big feelings even though we thought we'd just be excited. It's pretty normal to feel scared or nervous or even overexcited. How can I help you/support you when you feel those things? What can we have around to help you find some comfort?"

For Adolescents and Young Adults

"I thought I'd feel only excitement as the world opened up, but I'm learning there are a whole bunch of feelings that all the changes are going to bring. Given that we share space, I wanted to tell you this in

---

THINGS TO TRY: HELPING CHILDREN WITH OVERWHELMING FEELINGS AND TEACHING SELF-SOOTHING

- **Teach children the words for feelings**. Help children develop a vocabulary for their emotions. Model naming emotions by sharing your own when you have them. When you're coming to a red light you can say, "I feel so frustrated and impatient. I really didn't want to have to stop." Feeling word charts can be found all over the internet.
- **Use a family Zones of Regulation system**. Find and employ a color system that works for you so that each family member can easily express how they are in any given moment.
- **Create regulation stations**. Establish at least one small space dedicated to calming down and model using it. Leave a basket of objects that can help in this space (blanket, fidget toy, stuffed animal, etc.)
- **Brainstorm with your children about how to self-soothe without a screen**. Find non-digital ways of soothing, such as cuddling stuffed animals, singing together, deep breathing, working on a puzzle or handheld game, or other techniques.
- **Offer "muses" for your children to set their technology down for**. Leave embodied toys out for children to interact with. Skill toys (yo-yos, Kendamas, balance boards), playdough, building toys, and imaginary play items (such as play kitchens, doll houses, farms, etc.) all work.
- **Commit to small screen-free times**. Establish a ten- to sixty-minute block of time that the entire family goes without devices every day or week. Get buy in with the family and find alternate activities that everyone will enjoy.

case you experience me being on edge, and I also want to be a resource to you when you feel that way. Are there any things that we can put in place for you if you start feeling overwhelmed? I know it may sound dorky but I'm really serious. Can I help you carve out some space to just chill or help in any other way?"

PARENTING WITH EMPATHY AND RECOGNIZING CONFLICTING NEEDS

Many individual parents are going to have preferences and needs in going about re-entering embodied life that directly conflict with those of their children and, possibly, their partners. Exhausted and overwhelmed, some caretakers will be excited about the opening of schools for in-person instruction in a way that might inadvertently minimize their child's ambivalence and/or anxiety about returning. Other parents, who have enjoyed the extra togetherness of quarantine, may unconsciously hold their children back from much desired autonomy and freedom. The examples of the possible discrepancies could go on forever.

It'll be important for families to work together to name and work through the big feelings as well as the hopes they have for the future. Giving everyone a seat at this table will help families avoid the kind of backlash that can come with authoritative demands for children to simply comply with the needs and preferences of their parents. Rather than ordering our children around or offering unsolicited advice, it's going to be crucial that we create spaces where complex conversations can take place without fear of shame or need to shut down.

Metaphorically speaking, this is *not* a time to teach a child to swim by throwing them in the pool. With parents left stressed, stretched, and exhausted, there may be a strong reactive desire to scoot our children out the door more because we need the space than because they are ready for full-scale re-entry. Doing this puts our children at risk of denying the difficulties they've faced and/or any hesitance or concerns they have in re-engaging embodied life. When we do this, we can drive their concerns underground, where they fester into bigger mental health concerns.

It's important to realize that we are not fully out of the woods and there may be more shifting back and forth as we figure out how embodiment works post-pandemic. Our children will need flexibility and resilience as well as the ability to think critically as they are faced with these changes. We also must own the fact that the lack of social practice our children have had will have had a profound impact on the comfort levels of many children, youth, and young adults in returning to in-person life. It's likely they'll need safe places within to process this. They'll need help wading into the shallow end and then finding their courage to confront the deep end. This will take time and nuance and safe relationships with parents who respect them.

While it will be tempting to offer up our own parental insights as reliable and relevant truth about resuming communal living, it's crucial that we take an empathic stance with our children rather than an authoritarian one. Yes, we may *feel* as though we know what is best and right for our children in this time; we must, however, consider the bigger picture. If our goal is to raise children who can think critically and act responsibly, it's important that we practice parenting toward this end. We must find the right balance between offering instruction and listening and providing empathy as they make their way in the world. We're all going to make some mistakes. Ample grace will help.

It's important that we practice the art of asking open-ended questions that will help our child find their own answers. We must be available to the moment(s) when our child is feeling open to talk. This may not be on our schedule but will offer the most important conversations. Centering our communication with our children in empathy is the driving goal. When a child feels that their parent knows their real concerns and is rooting for them, they are more likely to have a strong sense of self and empathic connection to others.

To this end, every parent would be wise to practice the kind of skills involved in Motivational Interviewing (MI), which is described by its creator as, "a collaborative, goal-oriented style of communication . . . designed to strengthen personal motivation for and commitment to a specific goal by eliciting and exploring the person's own reasons for change within an atmosphere of acceptance and compassion." While MI was created for therapists and their work with clients, parents can benefit from learning about the kind of communication it privileges, which relies upon asking skillful questions to guide children to wise answers. This means having self-control in your conversations with your child, asking the kind of questions that inspire thought and nuance and letting your child work their way to the answers.[5]

An example could look like this: Child: "I hate school. I'm never going back!" Parent: "Oof . . . sounds rough. Are you angry, sad, overwhelmed?" Child: "No. I just hate it. Everybody's just being jerks." Parent: "In what ways?" Child: "I don't know, they just are." Parent: "It seems like you have really strong feelings when they act how they're acting. What are those feelings?" The goal is to help the child get to their own internal experience, where intervention and help can be offered rather than staying focused on the external things that they may not be able to change.

Along with the tools offered within the Motivational Interviewing approach, the following conversation starters can help get the ball rolling toward communicating deep compassion for our children.

For Young Children

"We've all had to do a lot of hard things this year. Now we're going to be doing a lot of new hard things. Sometimes I'm going to ask you to do things (go out in public, go to school) that I told you not to do before. It might feel confusing. Whenever you feel confused or angry or don't agree with me, you can tell me by saying, 'Can you explain that?' I'll try my best not to overreact when you do. Do you have any questions or ideas about this time?"

For Six- to Twelve-Year-Old Children

"How are you feeling about going back to school and being more out and about in the world? I'm guessing that sometimes, in the next weeks, we might disagree or I might do things that don't make sense. Do you have ideas of ways we can work through those times?"

For Adolescents and Young Adults

"How are you feeling about all the new decisions we're all going to need to be making about who to see and how to see them? (Stop to listen.) I am so thrilled for you that you'll get to return to a life that feels a bit more free. I'm guessing we're going to disagree about some things. I'd love for us to make a plan for how to interact when that happens. Maybe you can bring things up to me with enough time for me to mull over them and for us to have conversations that aren't too rushed or dramatic. What do you think?"

THINGS TO TRY: PARENTING WITH EMPATHY

- **Do a "What I need/What my kid needs" inventory.** Make lists of "What I need" and "What my kid needs" (e.g., I need time alone. I need time to work on home tasks. Jenn needs time with friends. Jenn needs help with fears). Draw lines between needs that conflict with each other (e.g., I need alone time and Jenn needs help with fears). Brainstorm about ways to address mismatches and identify resources and/or people that can help you navigate the difficult waters of tending to both sets of needs.
- **Practice the Platinum Rule.** "Treat others as *they* would like to be treated" is a hard, yet important, rule. It doesn't mean kowtowing or being subservient to your child but directs us to honor and respect our children's unique needs. If you're a verbal processor but your child isn't, find ways of offering conversations that come from their preferred methods.
- **Listen more, talk less.** Ask questions and let your child answer them before you offer wisdom, insight, or guidance.
- **Take steps to up your parental communication game.** Learn more about MI or nonviolent communication techniques and practice them outside of the heat of the moment.

DEPRESSION IN CHILDREN

All children feel sad from time to time, but some children experience levels of hopelessness and extreme sadness that indicate depression. When children are experiencing depression, they may show disinterest in things that used to bring them joy and might show signs of acting out or isolating rather than sadness. Children living with depression often have thoughts of wishing to die and, sometimes, consider suicide.

The CDC lists the following as examples of possible indicators of depression in children[6]:

- Feeling sad, hopeless, or irritable a lot of the time
- Not wanting to do or enjoy doing fun things
- Showing changes in eating patterns—eating a lot more or a lot less than usual
- Showing changes in sleep patterns—sleeping a lot more or a lot less than normal
- Showing changes in energy—being tired and sluggish or tense and restless a lot of the time
- Having a hard time paying attention
- Feeling worthless, useless, or guilty
- Showing self-injury and self-destructive behavior

This time of monotony and isolation will have taken a special toll on children, leading many of them to show signs of depression. This is important because childrens' developing brains rely upon new experiences and interpersonal interactions for robust wiring. When the brain doesn't have what it needs, it adapts by "pruning off" the regions not accessed. When this happens, a child may struggle more than normal as they lack the actual brainpower to cope. It's important to take inventory of these out-of-the-ordinary struggles and to discern if depression is settling in.

When a child is struggling with depression, life can feel very heavy for parents. This will be particularly true now as parents are living with less resilience than they might normally have, making it extra difficult to create the kind of spaces that children will need to be able to bear the weight of their suffering. Patience, grace, and help from others will be needed in large supply. If you feel that your child may be depressed, it's important to get them help from a professional with specific tools for treating pediatric depression. Too often parents wait to seek help, and this leaves the child at risk for their symptoms worsening.

It can be easy to over- or underreact when we suspect serious depression in our children. In reality, however, what is most needed is a steady and grounded approach to helping these children. They likely feel hopeless and scared by the depth of their feelings, and getting to be honest about these feelings with us will help them feel less alone. The reality is that depression responds well to treatment, and people who seek help typically learn to work through their experience and to thrive. When we stay calm and present and help them find resources, they are much more likely to get through the experience resiliently and with skills that will help them throughout their lives.

It's important to note, before leaving this topic, that children who suffered with depression prior to the pandemic may have enjoyed feeling more "normal" in this time of sheltering. In a world where they have felt highly different from other children who play and connect and seem to lead with joy, this time of cultural "depression" may have made them feel as though they fit in. If your child falls into this category, it's important to help them understand this experience.

Having conversations with our children about serious mental health concerns can feel particularly difficult. It's important, however, that we express a comfort in these kinds of talks so that our children can come to us, rather than relying upon less reliable helps, when they are hurting. Remembering that many parents wait too long to have these conversations, making things much worse, it's important to have them at the early stages of recognition of potential symptoms of depression. Here are some ideas of how to start these conversations.

For Children Up to Age Twelve

"It seems like some things are really bothering you these days. Does that feel true for you? Can you tell me a little bit (or draw or write) about how you feel? I love you so much and really want to help. I've noticed that you seem (insert words like tired, irritated, frustrated, sad) a lot of the time and those are pretty big feelings to hold. There are some people who help with that. I'd love for you and me to chat with one of them." The goal here is to set the stage for non-shaming and open conversations about their feelings and seeking help.

For Adolescents and Young Adults

"I know the world throws the word 'depression' around a lot and it means a gazillion things to different people. (Disclose, here, if you've ever experienced it personally or, if you haven't, if you have experience with it with others . . . humbly.) I've wondered if you might be experiencing it. If ever there was a time to be depressed, this is it. How is it for me to bring this up? (Listen well.) There are so many therapists who are compassionate and able to help. Would you be open to considering seeing one?" Then listen, listen, listen. If talking about it feels like too much, offer to take them for a drive or go for a walk. Sometimes them talking without having to look into your eyes might be easier. The goal is to open up a non-shaming and empathic conversation.

THINGS TO TRY: PARENTING CHILDREN WHO LIVE WITH DEPRESSION

- **Educate yourself.** Do some research (with credible sources such as the American Psychological Association) on the causes, signs, and symptoms of depression in children.
- **Don't under- or overreact.** Work out your own strong emotions apart from your child and then take a steady and loving approach to helping them find healing. Don't deny or become dramatic and get support for yourself rather than falling into despair. Your child will be taking in what you do and say and don't do and say, so working through your own reactions to their possible depression apart from them is important.
- **Offer normalizing and helpful resources for your family.** Educate yourself about the reality of depression in children and address any stigma you attach to it. Keep in mind that treatment is effective and accessible.
- **Help your child moderate their access to social media.** Social media use can cause depression. Help your child understand the way in which the curated images of others may impact their feelings about life and themselves. Do this in non-shaming ways.
- **Take mentions of suicidal thoughts seriously.** If you sense hints of suicidal risk or ideation, ask the child straight out about their thoughts and any plans. Those who are considering suicide or wishing to die are comforted by not having to keep those thoughts secret. If they do disclose these feelings or intentions, take them seriously and help them access a suicide prevention specialist (see the resource list in this book), therapist, or their pediatrician.

ANXIETY IN CHILDREN

When a child isn't able to outgrow fears in developmentally appropriate ways or when their concerns grow to the level of effecting their school, home, or play activities, a diagnosis of anxiety may be in order. Anxiety in children often manifests as irritability, worry, isolating behaviors, and/or anger, and children living with anxiety may have trouble sleeping and experience repeated head/stomach aches. Children who live with anxiety may have been dismissed and labeled "worriers" or may have shame about the behaviors that their anxiety fuels (such as tics or OCD rituals). For these reasons, children may not always express their fears verbally. They often suffer relatively alone.

The CDC lists the following as examples of anxiety in children[7]:

- Being very afraid when away from parents (separation anxiety)
- Having extreme fear about a specific thing or situation, such as dogs, insects, or going to the doctor (phobias)
- Being very afraid of school and other places where there are people (social anxiety)
- Being very worried about the future and about bad things happening (general anxiety)
- Having repeated episodes of sudden, unexpected, intense fear that come with symptoms like heart pounding, having trouble breathing, or feeling dizzy, shaky, or sweaty (panic disorder)

Relative to other mental health concerns, anxiety in all of its forms is of particular concern right now as COVID-19 necessitated real need for scrutiny about many parts of "normal" life. It's likely that anxiety has skyrocketed for children who lived with it prior to quarantine and that even children who aren't normally anxious have had spikes of nervousness as well. The need for intense care around physical connection with others, hand-washing, mask-wearing, and general moving about in the world coupled with the constant news of positive tests and lives lost to COVID-19 created hyper-awareness in children and adults that bordered on OCD. This is like gas to a flame for a person living with anxiety.

For parents with children who suffer with anxiety, this has likely been an exceedingly challenging time. Frequently adults see the "irrationality" of children's fears and feel impotent when they try to explain why the child doesn't need to be concerned. This often fuels a child's anxiety, which adds insult to the injury for parents who often can't realistically offer the child rational arguments disproving the likelihood of their fears coming to fruition.

What children need when they are feeling anxious is help getting through the anxiety, not help avoiding it. We must set realistic expectations and offer tools and support to empower them to do this. For example, if a child fears doctor visits it's best for a parent to express their understanding of the child's fear (e.g., "I hear how scared you are about your doctor's appointment") and then offer support in how to get through it (e.g., "I'll be there for you and we can do some things to help you get through the appointment together"). We can also work to avoid too much anticipation of experiences that trigger anxiety and also work with children to make plans about how to get through the experiences themselves.

It's best to practice ways of coping or make plans for how to work through anxieties when things aren't "hot" and the anxiousness isn't present. For instance, if going to school in person is raising fears for a child, we might begin talking about this in short bursts two or three days ahead, making a sequential plan for how the child can get through the experience even while anxious.

This might look like:

1. You can tell me when you want to talk about your fears leading up to that day. I will listen. I'll also do all I can to help you have fun things to do up until that day so your anxiety doesn't have as much opportunity to be front and center.
2. When you do feel anxious, we'll do some deep breathing together or we'll help you tap into your visualization of a safe space that you can go to in your mind.
3. On the morning of school, we'll have a great breakfast and we'll do some breathing on the way.

4. Once we're there, I'll give you a big hug that you can hold on to, and you can keep the little stone I'm giving you in your pocket to rub when you're worried.
5. At the end of the day, we'll go celebrate in whatever way you'd like.

During quarantine it's likely that our anxious children have witnessed anxiety in everyone around them and soaked that up as reinforcement for their fears. We have, consciously and unconsciously, modeled wonderful and horrible responses to our own anxieties, and children have internalized these. Now is the time to help them learn to get through their anxieties rather than avoiding them. In many ways our best approach will be to honor and recognize their anxiety and then offer opportunities to brainstorm about and onboard coping techniques. Just as we discussed earlier, children who are anxious wish for an intense response from their parents . . . a response that matches the intensity of their fears. If escalating their anxiety gets a stronger response from you, they're likely to do that, often without consciously knowing that is what they're doing. If, however, they feel intensely responded to (e.g., "Wow. I hear how scared you are about this. I feel you. It's super real to you. I can also tell you that I think your feelings may be driving your thoughts. Let's work together to help you with the feelings"), they may be more able to let some of their anxiety go.

As with concerns about depression, parents who have a lingering feeling that their child may be living with anxiety would be best served by seeking help sooner rather than later. Anxiety, like depression, is very responsive to treatment, and children who suffer from it can recover. It is difficult to do the work of finding the right help, but the payoff will be huge.

To talk with children about getting help dealing with their anxieties, the following starters may offer help.

For Young Children

"I notice that you feel worried a lot of the time. There are some things that we could do and people that we could talk with that might be able to help you learn how to cope. I'd love to learn with you so I can help you."

For Six- to Twelve-Year-Old Children

"Worry can really zap people of feeling like they can be okay. I've noticed that your worry seems to be dragging you down these days. Am I right about that? What does it feel like in your body and mind? Are there things you do when you feel nervous? Do you feel safe telling me about them? I know of someone who could help us learn how to help you with these things."

For Adolescents and Young Adults

"I know that most of the world feels anxious right now. I know I do. I've wondered if some of the things I notice these days are because you feel that way. Do you? How is your anxiety impacting you these days? Are you finding yourself doing certain things to try to control or contain it? I want to be a safe person and I also don't have what is needed in the way of knowledge or tools. I'd really love to help you/us find those."

FOSTERING RESILIENCE DURING RE-ENTRY

Resilience refers to the ability to face distress without experiencing significant psychological dysfunction or symptoms. The Center on the Developing Child at Harvard University describes resilience in this way: "One way to understand the development of resilience is to visualize a balance scale or seesaw. Protective experiences and coping skills on one side counterbalance significant adversity on the other. Resilience is evident when a child's health and development tips toward positive outcomes—even when a heavy load of factors is stacked on the negative outcome side."[8] In essence, resilience involves the capability to "bounce back" after significant hardship, failure, disappointment, and, in some cases, trauma.

THINGS TO TRY: PARENTING CHILDREN
WHO LIVE WITH ANXIETY

- **Don't expect to eliminate anxiety and instead focus on onboarding tools for getting through it.** Don't intentionally expose your child to anxiety but don't avoid things that make them anxious. Learn some tools for managing it (breathing, redirecting attention, physical interventions) and teach them to your child.
- **Respect your child and set realistic expectations.** We can't promise our children that their anxieties aren't "real" or "likely." We must connect and offer specific tools without inflating or denying their fears.
- **Make plans with your child outside of the anxious moment.** Set out a step-by-step scheme that the two of you will use to get through and practice it. Make a picture or word chart to help with this.
- **Help your child moderate their access to social media.** Social media use can cause anxiety. Help your child step away from it when they are suffering with anxiety. Direct them to physically active outlets or to slow-moving, high-quality digital content instead.
- **Seek expert help if needed.** If your child's anxiety is severe and/or manifesting in tics, social anxiety preventing them from interacting with the world, phobias, or panic, then work with your pediatrician to find a therapist who specializes in the treatment of pediatric anxiety. While many licensed general practitioners offer treatment for anxiety, it is best for this particular malady to work with someone with special training in gold-standard anxiety treatments for children.

Kids who have a foundation of resilience are more likely to tolerate discomfort and to take appropriate risks. They don't automatically think that being uncomfortable means that they should "jump ship" and are able to experience failures as tools for learning and growth. Social worker Katie Hurley rightly says, "Kids need to experience discomfort so that they can learn to work through it and develop their own problem-solving skills. Without this skill-set in place, kids will experience anxiety and shut down in the face of adversity."[9]

It's appropriate to assume that most children's lives have been stacked with significant levels of adversity and discomfort during the course of the pandemic. Some children will have found ways of working with the adversity and continuing to thrive while others may be facing a year's worth of accumulated discomfort in a way that has led to real psychological concerns. Children who had flagging levels of resilience before quarantine are likely to feel even less of it now.

Faced with severely restricted experiential opportunities during quarantine, it's likely that entire family sets are experiencing low resilience levels at the end of quarantine since positive experiences (to counterbalance the difficult ones) are required for building and maintaining the ability to cope and bounce back. This is important because, if a child has at least one relationship with a supportive adult and has been taught some adaptive skills around how to regulate oneself (self-soothing, naming and working through feelings, etc.), they are in a position to continue developing their resilience even when faced with hardship. For this reason, we must focus on enhancing the resilience of the adults in a child's life since children who show resilience aren't starkly self-reliant, but instead feel capable of asking for help when they need it.

This means that it's more important than ever to offer ourselves to children as safe people who will listen, respect them, and offer help for them to work through their own difficulties rather than avoid them. It's not up to us to rescue them. It is up to us to get in the trenches with them and help them find solutions to the difficulties that they face.

Many of the skills referenced in the previous "Things to Try" sections apply here. Showing up and communicating that you see and "feel" the children in your life, leading with empathy, listening attentively, asking the child to help you know what they need, supporting rather than rescuing, and offering tools rather than answers will all help the children in our lives develop the kind of resiliency that will help them recover and find redeeming elements about the hardships that COVID-19 has presented them with.

For Young Children

"Things have been so weird this year. It might feel really scary to start being with more people and to change how we do things. I want to help you find some ways of telling me when you feel afraid or feel funky feelings. If you can't find words you can just say, 'The weird feelings are here,' and I'll help you figure out how to work them out."

For Middle-Age Children

"While we've been doing online school and everything else online, you might have lost some of the comfort of being with others. It might be weird and super uncomfortable to be back with people or you might have a hard time containing your excitement. If that happens, let me know. I'd love to help you figure out some things you can do to work through your feelings."

For Adolescents and Young Adults

"This year has taken a hit to my ability to work things through in the ways I normally would. I sort of feel 'on edge.' Can you relate? I'm thinking that that's likely to get worse as things change again and we are trying to figure out how to re-enter the world. If there are things that I can offer to help you work through any feelings you have, I'm here for you. I'll try not to bug you and let you ask . . . just know that I'm here."

HOW TO GET PROFESSIONAL HELP WHEN/IF YOU NEED IT

There are going to be families who will benefit from a "coach" or helper in this time. Family therapists as well as individual therapists who work with specific age demographics can be very helpful. While a parent may have to overcome some of the stigma associated with asking for such help, it's important to reframe the process. Rather than thinking, "Only messed-up families need to go to therapy," an accurate reframe, such as, "Strong families are capable of doing hard things in order to get through challenges healthily," can empower everyone involved.

Most of us would never hope to fix complicated issues related to our cars. When our devices give us grief, we are quick to head to the Genius Bar. When our children experience physical symptoms, we take them to the doctor without doubting ourselves. Reaching out to mental health professionals to help our children and families soar is no different. These helpers are driven by compassion and have special skills to help

---

THINGS TO TRY: FOSTERING RESILIENCY IN CHILDREN

- **Consider your own level of comfort with failure and discomfort.** Identify ways in which your own modeling may cause your child to fear appropriate risks and/or failures.
- **Make yourself aware of your "rescuing" behaviors and triggers.** Take stock of the situations that your child may find themselves in that prompt a "rescue" response in you. Make a plan for letting this go in order to encourage healthy risk-taking to build resilience.
- **Identify where you take over for your child rather than empowering them.** Identify the passive and active ways that you problem-solve for your child rather than with them and determine to stop. Direct your attention, instead, to helping them work through the problem.
- **Get comfortable with new ways of thinking about failure and success.** Challenge your own notion of both and daydream about how a resilient child is different or the same as a successful one. Set your sites on raising a resilient child.
- **Reassess self-soothing skills.** Being able to soothe oneself when distressed is a huge benefit when developing resilience. Help your child and yourself identify personally soothing routines and make them easily accessible.

---

us move through challenging situations. We've gotten through a global pandemic . . . why not use every tool available to help us thrive from here? Here are some ways to start conversations about getting help with our children.

For Children Up to Age Twelve

"I've been noticing some things that don't seem 'normal' for you. Do you have any big experiences or feelings you want to talk about? Sometimes these kinds of things can stay stuck in our minds/hearts in a way that makes it hard to do the normal things we do in a day. There's a cool place where you can go and play/talk with a person who really understands children, and they can help you talk about anything you're feeling. I've made us an appointment to try it out and see how it feels."

For Adolescents and Young Adults

"I'm guessing you know all about therapy and probably have friends who go. Is that true? I want to be really gentle in asking you if you might want to/be willing to give therapy a try. This is really hard to bring up because I don't want you to feel like I'm telling you that there's something wrong with you or that I'm not here for you. I'm just realizing that I don't have everything you might need to weather this time well. How is this feeling to you?"

THINGS TO TRY: GETTING HELP FOR YOUR CHILD

To find the right professional for your family, ask your physician or your child's pediatrician, school counselors, or friends for names of family therapists or individual therapists in your area. Do your research and ask if you can have a fifteen-minute phone call with the therapist you choose to assess if their temperament seems as though it will vibe with your family. If your child is younger than sixteen and would benefit from individual therapy, there is a good chance that the therapist will make a plan with you of how and what they will share information with you. If older than sixteen, they'll discuss the ways in which confidentiality supports and helps the therapy process and talk you through what to do if and when you are concerned.

The kinds of therapy you might consider would be family therapy (done with all parties in the room), individual talk therapy (for your child, yourself and your partner, or for an older child), play therapy (for children through middle elementary school), or occupational therapy if your child is facing physiological fallout from the pandemic (tics or movement habits, difficulty sitting, etc.).

Remember, these professionals do this work because of their deep and informed care for others. Their goal is to help you all thrive.

III

RESTARTING—WHAT WE CAN DO TO EASE OUR TRANSITION

6

HABITS AND NORMS

Cass entered pandemic life as a working nurse in graduate school. As soon as cases began to surge, she supported the work of the emerging COVID-19 unit as well as doing formal and informal community outreach to educate people about the virus. When civil unrest unfolded around issues of race, she donned a gas mask and served as a medic at protests. All the while, graduate school assignments kept coming in. Over time, Cass found herself near burnout trying to live into the value of being responsive and responsible. Her strong values wouldn't allow her to simply "drop" things, but her habit of doing the right thing as much and as often as possible became unsustainable. Whenever a new need arose, she'd offer to fill it, realizing, too late, that she may not have the energy it would require to fill with the excellence she demanded of herself.

Mo, a respected and connected young adult, entered pandemic in a high-tech job that lacked meaning but supported his economic needs. When his employer transitioned all work to home, Mo struggled with loneliness and feelings of isolation. Overwhelmed by these feelings, he dove deep into a MMOG (Massively Multiplayer Online Gaming) platform as a way of feeling like part of a team. Over time, he began logging in to play at lunch and, again, as soon as his workday ended. As the weeks of isolation turned into months of it, Mo began to feel an increasing drive to engage his gaming community alongside an ambient sense of unrest and loneliness. He yearned for a friend who he could see, even at a distance, and people who he could interact with in the sphere of his embodied life. Even though these feelings troubled him, his urge to spend his free time in the game universe was so strong that he couldn't resist. He was grateful for these online relationships but lonely when he was outside of the game space.

Both Cass's and Mo's stories are all too familiar. Life hands us circumstances and we immediately fall into habits that help us get by. We're faced with a challenge and we react, our knee-jerk behaviors becoming the beginning of habits that will get us through. For Cass, the habit of stepping up in seemingly "responsible" ways without considering the cost is a habit that drives her behaviors and ends up hurting her mental health. For Mo, habitually turning to connections in the digital domain means little time for, and thereby a passive denial of the need or desire for, embodied forms of engagement. This habit impacts his sense of well-being and the options he can see to address his dissatisfaction.

WHAT ARE HABITS?

Habits are behavioral patterns that we engage largely without thinking. They're often reflexive, and we act on them automatically. Habits quickly become handholds of sorts, and it's common to feel that we can't get by without them. In many ways we feel as though they make our lives "easier" as they give us behaviors to engage without thinking. We drink coffee when we feel tired, bite our nails when we're anxious, laugh nervously when we don't know what to say, and engage a million other habits that feel as though they help us cope. The problem is that our habits have as much potential to hurt or prevent our health as they do to help us.

This is not just a behavioral reality since our habits are supported by repetitive actions that have their roots in the brain. In later chapters we'll explore this more but, for now, it's important to understand that when we are exhausted and emotionally overwhelmed, we're likely to be especially vulnerable to letting our habits run the show. It feels like too much work to break them, and alternatives are hard to imagine.

Mo's habit of connecting with people in online spaces consumes a lot of his free time. It has also left him in a space where he can have a significant amount of surface connection with many people in one relatively low-stakes space, protected by the anonymity of the game. This not only robs him of time to invest in embodied relationships but also impacts his actual ability to take communication and relational risks with embodied others. Without practicing those behaviors, he will not maintain the neural wiring necessary to help him do these things, causing him to lack confidence in them and avoid them. This mix of reinforcement for both behavioral habits and those rooted in the brain is a powerful one, impacting Mo at surface and deep levels.

Along with Mo, as we've been stuck in the reality of highly repetitive lives with very little external stimuli apart from that which we find in digital spaces, our habits related to our technology use are particularly strong and entrenched, impacting our embodied relationships with ourselves and others. Our devices have offered us unlimited freedom in a time where we've had relatively none of it in our embodied lives. It makes sense that our habits dictated that we turn to all things digital for help getting through this difficult time.

WHAT ARE NORMS?

The word "norm" is a reference to normative behaviors or patterns of action. Norms serve as more conscious and intentionally chosen alternatives to habits and can serve to help us break maladaptive cycles of behavior. Norms help direct our behavior, thoughts, and feelings and are built upon chosen or aspirational values or goals. In many ways, norms offer us healthier alternatives to responding to life than habits do in that they are rooted in something more than reflexive reactions.

If Cass habitually responds to requests for her service with "yes," she will remain at risk of overwhelm and burnout. In many ways, her unconscious, reflexive "yes" to the world means she'll likely have to say "no" to self-care. If, instead, she taps into her deep value for responsibility by setting the norm of taking a break between being asked to do something and saying yes or no, she will have the space and reflective time that might empower her to make different and more wholistic choices about remaining healthy while also serving her community.

BREAKING HABITS

Where habits form, many behavioral patterns grow roots to support them as there are many small habits that support the larger ones. Let's say we have a habit of smoking. The small behaviors that support this action can easily, over time, become as rewarding as the behavior itself. The opportunities to take a break from the situations we are in and go outside and draw deep breaths for a period of time quickly become rewarding for those who smoke. The same is true for many of our habits. We scroll through social media and are rewarded by feeling caught up on trends and "connected" to those we follow. Never mind if our scrolling also causes us to feel depressed and lacking. We binge on a game for a day, turning to it in every free minute, and are rewarded by unlocking new levels and feeling a sense of expansiveness even if it means ignoring our sense of agitation and dysregulation from obsessively grabbing for our game in all free moments.

Frequently, rewards keep us from realizing the costs that come with our habits. Our smoking leaves us with increased health risks, our scrolling robs us of the opportunity to develop our in-person relational and social skills, and our time spent with online games takes us away from engaging our physical bodies. Ask anyone facing an entrenched habit if they'd rather try to stop or never have started the behavior, and many will say they wish they'd never started. Habits are very, very hard to break.

Trying to break habits cold turkey is especially difficult. While some people are successful in simply stopping in this way, most of us will be benefitted by making a plan to extinguish the habit in more gradual ways. For changes to be lasting, we are benefitted by tending to the emotional realities we will face as we try to break our habits. Making a clear and cohesive plan will help with this as well as with the creation of new norms.

Wiring the brain for certain types of actions and reactions (habits) does double duty. When we create habits, we actively neglect exposing our brains to experiences that could create proficiency toward other possibilities. When we practice smoking, we actively rob the brain of the opportunity to seek out soothing experiences that do not involve cigarettes. When we reflexively reach for our devices whenever we have a free moment, we forego the experience of forcing a pause in our action, giving ourselves the opportunity to select our next action from a variety of choices.

There are ways, however, to allow ourselves to feel uncomfortable and gain a new skill, in order to break our habits, and many small actions can help with this. Instead of scrolling our way through the time spent at a red light, perhaps we could write a limerick in our mind instead. Rather than bingeing on YouTube during a break at work, we could stand up and take a walk around our office or go outside for a breath or two of fresh air. Whenever we pause between our reflex to act in a habitual way, we offer our brains and bodies the opportunity to develop new expertise. This is crucial for success in breaking those habits that passively harm us or that no longer serve us.

COVID-19 and the resulting safety protocols it ushered in have offered opportunities for us to create many habits. We've begun sleeping in and rolling out of bed to meetings. We've set aside personal hygiene. We've, out of necessity, had everything delivered to us, thereby losing practice at benign social interactions at the store, laundromat, or bakery. Out of our emotional and physical exhaustion, we may have developed communication habits with our families and friends that may be overly dismissive, confrontational, or avoidant. Every part of our lives has likely been affected by habits we've fallen into in this very stressful time.

If we hope to stay healthy as we experience increasing opportunities to engage the wide world around us, we'll need to offer ourselves help and support in breaking many of our habits. Learning to enact a pause between impulse and action is the first, and possibly, most difficult task to onboard, but doing so will allow us to create healthy post-pandemic norms.

HOW TO PAUSE IN THE FACE OF A HABIT

Many of us are familiar with the concept of classical conditioning as explained by Russian researcher Ivan Pavlov's famous salivation experiment. In his work, Pavlov found that there were certain behaviors that dogs didn't need to learn, such as salivating when they were exposed to food, but that there were some behaviors that could be effectively taught. To prove this, Pavlov inserted small test tubes into dogs' cheeks to measure their salivation when they were fed powdered meat. Once he measured the amount of saliva produced with the sight and smell of meat, he introduced the sound of a metronome shortly before the meat was presented and, over time, the metronome alone caused a salivation response in the dogs even when food wasn't produced. In essence, the dogs learned to salivate when the metronome was presented because they knew that food was coming. Once entrenched, this response became a habituated one that required no reward.

We see similar forms of classical conditioning in our engagement with our technology, and these responses were only intensified by quarantine. Devoid of most forms of casual interaction, our devices became increasingly important to us. They offered us connection, entertainment, and opportunities to work and learn. Just as dogs began to respond to what was originally a precursor to food, we've come to respond to our devices in the way we might have previously only responded to embodied interactions and experiences.

Perhaps the most entrenched habits we'll need to break post-pandemic will be those related to the way in which we engage our technology. It's become our portal into almost everything for the past year, and we've grown even more highly dependent upon our devices than we were prior to quarantine. Even still, excessive use can cause mental health difficulties, make us irritated and agitated, emotionally dysregulate us, and lead us to feel less capable in and of ourselves.

It seems, then, that the goal is to condition ourselves to take a pause between our impulse to grab our devices and the behavioral act of doing so. We are benefitted by learning and practicing this kind of hearty pause as it enhances our feelings of using our personal agency well. Creating this pause, however, will not be easy. It's much easier to follow our impulsive gut and grab for our phone. It's much harder to introduce the idea of a pause in this unconscious process.

This is particularly important now because, as the world opens, it's likely that we'll feel hesitant, at times, to re-engage interpersonal interactions. Given the severe drop-off of experience with in-person encounters, reconnecting socially may be anxiety producing. This might be related to a lack of experience while sheltering in place and also to any preexisting habits that have been made more intractable in this time of isolation. If we weren't comfortable with in-person connection prior to the pandemic, this has likely been made worse. If we felt insecure about our verbal communication skills, we probably feel even more so now. Being able to take a pause and determine how we might help ourselves through the awkward steps of trying to become socially engaged again will be a huge boon to our health and happiness.

What this requires is pairing our impulse to reach for our devices with something other than the immediate rewards that they provide. Practically, this means becoming aware of the kind of feelings and experiences that drive us to grab for our phones and then learning to pause when we are in these spaces. For example, when we feel lonely and tempted to reach for our phone to text people, pausing might allow us to determine that a five-minute phone call would be a better, richer option than texting, also helping us readjust to voice-to-voice conversations. Or, pairing a pause with the experience of the gnawing and ambient feeling of missing out could help us determine that a different activity would be better for us than doomscrolling.

There is huge power in pausing, and this can be harnessed by continually imagining a pause in our behavior between impulse and action. Committing to do this is the first, and most potent, start to breaking habits that don't serve us.

SETTING NORMS

While it's easier to establish healthy norms than it is to break bad habits, it is rarely easy to set norms. Since the kind of norms that would lead to happier and healthier lives are intimately connected to our values, these can be a tool for helping us determine what norms will be most potent for us to pursue.

Core values lists and assessment tools can be found easily online. For our purposes here, however, we can work from this list of common values culled from research[1]:

Authenticity	Achievement	Adventure	Authority
Autonomy	Balance	Beauty	Boldness
Compassion	Challenge	Citizenship	Community
Competency	Contribution	Creativity	Curiosity
Determination	Fairness	Faith	Fame
Friendships	Fun	Growth	Happiness
Honesty	Humor	Influence	Inner Harmony
Justice	Kindness	Knowledge	Leadership
Learning	Love	Loyalty	Meaningful Work
Openness	Optimism	Peace	Pleasure
Poise	Popularity	Recognition	Religion
Reputation	Respect	Responsibility	Security
Self-Respect	Service	Spirituality	Stability
Success	Status	Trustworthiness	Wealth
Wisdom			

In looking over this list, identifying no more than five core values can help us to determine where to focus our norm-setting efforts. Once we have identified the five values most important to us, we can begin considering which of our habits help us live in concert with these values or not.

As an example, let's consider the life of Tobi, who values competency, curiosity, and love above all else. With his work as a nurse taking much of his time and energy, Tobi finds that engaging in hobbies helps him maintain balance. With the power of the internet at his fingers, he pursues all kinds of possible activities that interest him and procures everything he needs to make his efforts successful. He spends hours researching the skills and tools necessary for rock climbing and cooking and orders everything required for getting started. He checks regularly for new information and spends inordinate amounts of time reading everything he can on how to succeed in his new endeavors.

While Tobi's habit of staying on top of all relevant information may feel as though it checks the boxes of being true to his curiosity and assuring his competence, the reality is that he isn't actually living in to either value. If he were, instead, to get outside or in the kitchen, experimenting with climbing techniques or recipes and ingredients, he would be living out of curiosity and in to finding personal competence.

Most of us can relate to this story. Our habits help us feel as though we are exploring and expressing our values but, in reality, we are often only engaging them in very intellectual and one-dimensional ways. It's crucial, as we re-enter embodied living and re-orient our relationship to ourself, technology, and others, that we take intentional steps to use our values as spotting points in norm setting.

SETTING NORMS

- **Do a values assessment**. Write each of the values in the list provided in this chapter on a sticky note and put them on a door, wall, or window. Sort them according to "Highly Resonate," "Resonate Some," and "Aren't at the Top of My List" categories. Remove all but the "Highly Resonate" category and prioritize them from most to least resonant. Remove all but the top five and, for a day or two, live as though these are the only values you ascribe to. The goal isn't perfection, it's getting a good-enough mix to work from.
- **Do a values/habits comparison**. Consider two or three of your habits that dominate your day-to-day life. Hold each of them up to each of your top five values. Can each value and habit live in the presence of each other or do they conflict? Identify several steps you could take to stop any habits that are mismatched with your values.

SPOTTING POINTS

If we have some pre-pandemic or during-pandemic habits to break, now is the perfect time to get to it. As we do so, we can put in place some post-pandemic norms to support our success. Spotting points can help us as we do the hard work of letting go of our habits in order to live in accordance with our value-driven norms.

Spotting points are tools that can be used to help us stop maladaptive habits, change our behaviors, and meet goals. Dancers often choose a point out on the horizon to fix their gaze on. Once they have locked their sight on the point, they can continue to "spot" toward it by keeping their vision tuned to it at all times. When they are spinning, they snap their head around to keep their spotting point in sight, and the same is true as they move across the stage. Snowboarders and skiers use chalk marks in the snow as spotting points, and golfers use flags emerging from holes.

Our values, in essence, can serve as spotting points for norm setting. If we value connection, we can set norms that ensure we'll connect with ourselves and others in ways that are meaningful. If we chose stability as a core value, that can serve as a spotting point directing the choices we make about where and how we spend our time and energy.

There is a difference between the spotting points that culture, our parents/pasts, vocations, and interests might dictate for us and the ones that we, ourselves, choose. It is crucial that we know the difference and make intentional choices about which points/values we will spot to. For example, culture might dictate that a woman should spot toward nurturing or supportive trajectories whereas any individual woman may feel a deep sense of calling to any other number of spotting points. The importance of every person being free to live into their own unique desires, gifts, and callings is paramount, as is the reality that culture has brought many people to this current moment bereft of the opportunities that others have had. It's up to each of us to work toward addressing inequities in the world that privilege some and oppress others. While we work to set our own spotting points, we must commit to empowering all people to have the freedom and resources necessary to spot to their own values and callings.

To get a visceral feeling of what it's like for our minds to spot to unclarified spotting points, we might do the following experiment. Using sticky notes, list expectations you feel from others or yourself (e.g., to be perfect, to get everything done, to respond to all communication, etc.), roles that you fill in the world (e.g., employee, student, friend, partner, community member, parent, etc.), and the goals you are working toward (e.g., cleaning out all clutter, finishing a major personal or professional project, using technology less, etc.). Write one value, role, or goal per sticky note. Place these randomly around the walls in a room. Identify one that you want to "spot toward." Really look at it and identify where it is and then start spinning. See how hard it is to keep your focus on it when there are so many other identical objects on the walls around it. This is your life without a whittling down of the norms and goals that can really help you stay focused.

To go further, we can check the way in which our values are guiding our norm-setting efforts by trying this experiment. With five sheets of paper, write one value from your earlier assessment on each of them, large enough that you can see them from a bit of a distance. Tape them on the walls around a room so that, by standing in the center of the room, you can turn in a circle and see each one at a time (roughly). Spin slowly a few times, noticing each of the values as a spotting point. Identify a habit that is not serving you and identify the feelings, experiences, and thoughts that sustain it. Now, with that habit in mind, turn toward each of the values on the walls and identify how spotting toward that particular value might help you identify behaviors or experiences to engage instead that would help extinguish the habit. Also work to identify actions, feelings, or thoughts emanating from each value that would be better spotting points than the habit itself.

When we know where we're going, it's much easier to make a plan to get there. For that reason, getting specific about what we're spotting to can be the thing that makes or breaks our ability to set norms that will help us thrive.

IDENTIFYING VALUES AND SETTING SPOTTING POINTS

- **Experiment with spotting points.** Try the exercise with sticky notes, experimenting with a new spotting point each day until you find your most fitting one.
- **Use your values to guide your norm-setting process.** Review your top values as identified earlier in the chapter and assess their goodness of fit with each of your top spotting points.

PREPARING THE ENVIRONMENT FOR HABIT BREAKING AND NORM SETTING

Breaking habits is never easy. This is especially true when we are tired and overwhelmed. As the world re-opens, we will all be facing a myriad of feelings. Relief, grief, anxiety, anger, joy, and more will all be surging through our systems. For this reason, it's important to make actual plans about what habits we hope to break and what norms we'd like to put in their places. This will help us to set the stage for success. This may look like reworking our schedules so that it isn't as easy to fall into habits or like asking some trusted others to ask us how we're doing. It might involve stocking our homes with embodied offerings that keep us from habitually reaching for our devices or our calendars with activities that will engage our whole bodies and minds. Without this kind of advance preparation, we are less likely to succeed.

PREPARING THE ENVIRONMENT FOR YOUR EFFORTS

- **Take note of the ways that your surroundings support your habits rather than encouraging new behaviors**. Moving from room to room in your house, identify the things that support you falling into or maintaining the habits. Is your furniture arranged in such a way that it's oriented toward screens rather than windows or other focal points of beauty?
- **Move some things around**. With what you noticed from the above exercise in mind, make changes that will move you from habitual living (e.g., plopping down on the couch, grabbing the remote, and turning on the TV) to living from healthy norms (e.g., with the couch now facing the window and the remote living in a basket near the TV, you plop down and grab a bowl of kinetic sand and interact with it for at least ten minutes before engaging the flat screen).
- **Ask for help, support, and accountability**. Going through the contacts in your phone, identify two or three people who you can share your goals with and schedule a time to do so. If appropriate, ask them if they might check in on your progress periodically, and offer to do the same for them.

7

BEGINNER'S MIND

We all know stories about people who lost fifty pounds while eating only bacon and cheese but then gained it all back when they expanded their food choices. Similarly, we try to take technology fasts and end up bingeing when we pick up our devices again. We participate in No Alcohol January only to overconsume in February. We've had a year of forced social isolation and we're chomping at the bit to re-engage. It's likely that we won't be perfect at doing that in safe and measured ways. Humans are not necessarily naturally suited to creating healthy and sustainable habits. Moderation doesn't seem to be our jam.

After months of imposed restrictions due to the pandemic, it's likely we'll be wildly excited about returning to the embodied life of hugging, coffee dates, and social outings. We have potent memories of what freedom in our movements felt like and plentiful plans about everything we can't wait to resume. In many ways, that's all we've been thinking about: resuming life as we knew it.

At the same time, because the quarantine was so intense, we probably feel an ambient sense of concern and awkwardness in reconnecting with our communities in person. Adding complexity to the mix, many of us have come to appreciate some of the gifts inherent in this time of quarantining and social distancing. Participating in meetings from our recliners, having to be dressed only from the waist up, foregoing bras, and having sanctioned breaks from social gatherings or family-get-together drama are just a few of the perks some will miss as we re-open communal living. In many ways, this time has offered us gifts.

As we've learned, habits are hard to break. Even so, new beginnings offer an especially powerful opportunity to establish new norms. Given that we entered pandemic-related quarantine with habits that were harmful, this new global beginning may be just what we need to motivate us to engage with ourselves, our bodies, the important other in our lives, and devices in healthier ways.

HABITS AND SHOSHIN: BEGINNER'S MIND

Most of us have habits dictating the way we live our lives. Pre-pandemic, we had habitual ways that we interacted with ourselves and the world. When the world shut down, everything ground to a halt with the exception of the memory of what we longed for: freedom and in-person connection. While we dealt with feelings of isolation and restriction, these powerful memories turned to passionate longings we hold on to today.

If, pre-pandemic, we had managed to create relatively healthy routines for our internal lives, health, and interpersonal relationships, these pandemic longings may result in an unexamined re-entry that is smooth and healthful. If, however, we've come to idealize pre-pandemic routines that weren't necessarily healthy, we are likely to revert to habits that hurt us. We might also rush into new, unconsidered habits that aren't optimal for our well-being.

HOW OUR HABITS CAME TO BE

We began creating patterns of living the moment we were born and continue developing them every day. The parents we were (or were not) raised by, the safety (or lack thereof) of our primary environments, and the opportunities we were given (or robbed of)—as well as a million other variables—subtly impact the way we interact with our internal and external worlds. Our daily routines are the result of unlimited lessons we've passively and actively learned from people, experiences, and feelings. By the time we reach adulthood, these lessons have formed themselves into behavioral, emotional, and intellectual patterns that define and direct us. The patterns dictating our daily life may well be predicated on lessons learned that are not gold standard, tested, and evaluated for optimal well-being. This means that, unless we're discerning about the lessons we're learning, we may continue living life in patterns that hurt us.

Author James Clear brilliantly states, "Here's the hard question: Who is to say that the way you originally learned something is the best way? What if you simply learned *one* way of doing things, not *the* way of doing things? . . . Who is to say that the way you originally learned a skill is the best way? Most people think they are experts in a field, but they are really just experts in a particular style. . . . In this way, we become a slave to our old beliefs without even realizing it. We adopt a philosophy or strategy based on what we have been exposed to, without knowing if it's the optimal way to do things."[1] If living a healthy life is a skill, then we'd do well to heed Clear's wisdom. Perhaps this is the perfect time to realize we've learned *one* way of living, not *the* way of living healthy.

Each of us is the expert of our own life—whether that life be thriving, dormant, or entirely neglected. Even when we'd like to change certain things about our day-to-day life, we tend to be driven by unconscious biases that lead us to believe that our choices and actions, thoughts and feelings, beliefs and values are the "right" ones. We search our surroundings for evidence that this is true and do mental gymnastics to keep it so. Our selective attention leads us to disregard information or experiences that may challenge our internal systems of belief. In other words, if we *believe* we are wise or *see* ourselves as smart, we may disregard evidence that contradicts this belief and hyper-attend to evidence that supports it. Similarly, we seek out information that confirms our assumptions and keep away from information that might challenge our beliefs, behaviors, biases, and assumptions. This is called "confirmation bias."

Selective attention and confirmation bias mix potently with our habits to make us certain that our skill at life-living is that of an expert. Interestingly, quarantine life has left us space to idealize the lives we lived before the pandemic and to disregard or downplay insights or recollections about the parts of our daily patterns that may have been less than healthful. Without reflecting on this reality, we are all chomping at the bit to return to life "as normal" and as experts of our own normalcy. This disregards the incredible potential that our new global embodied opportunities and freedoms might lead us to—with healthy reflection and planning.

It is crucial to note that, whether we take ownership of where we are in life or feel a victim of it, there are active systemic forces that have shaped our realities. Racism, classism, sexism, ageism, ableism, homophobia, and the effects of colonialization are all very real forces that profoundly shape daily life for many. For those who live within specific beautiful and vulnerable demographic groups, many obstacles have been erected by society to undermine and deter their efforts and choices. For these individuals and communities, the continual restriction and withholding of resources and opportunities has offered them less control over their daily choices and, thus, life trajectories. As we transition to speaking of beginner's mind, we need to realize that it is an evidence of privilege when we are able to easily step back and make choices about changing patterns and habits. May we all do our part to work toward *all* people living in such a way that they can access this human right.

APPROACHING THE NEW REALITY WITH SHOSHIN

There is a practice, born in Zen Buddhism, referred to as Shosin, which encourages seeing life and all things in it from a "beginner's mind." Approaching situations from this mindset means dropping all assumptions and expectations about how or what something *might be* and, instead, seeing *what is* with openness and curiosity. This requires adopting a posture akin to that of a child who looks at the world with wonder. It means vulnerably addressing situations from the perspective of a learner who isn't invested in specific outcomes, but, rather, in exploring all potential.

We've all had interactions with people who believe they are an expert in all things. They know more than we know, they've experienced more than we could ever experience, and they are insufferable in keeping us aware of these realities. Nobody wants to be this person, yet, when it comes to each of our lives, we all live this way.

When people are *the* expert, they know a lot. They rely heavily on this knowledge and on the feeling of security that mastery offers them. They have many attitudes, opinions, and ideas that they (consciously or unconsciously) believe to be "gold standard." They talk more and listen less, teach more and learn less. These folks feel, and communicate to the world, that they "have arrived."

When people are beginners, they know very little. Every option seems like a possible one. Curiosity rules the day. Starting something new requires learning and an open mind about how to proceed. Being at the genesis of something requires a heart open to potential and all that is possible. These are all gifts of being a beginner. To quote Zen master Shunryu Suzuki, "In the beginner's mind there are many possibilities, but in the expert's there are few."[2]

The benefits of engaging tasks from a beginner's mind are many and summarized well by Leo Babauta on ZenHabits.net:

- **Better experiences**. You aren't clouded by prejudgments, pre-conceptions, fantasies about what it should be, or assumptions about how you already know it will be. When you don't have these, you can't be disappointed or frustrated by the experience, because there's no fantasy or preconception to compare it to.
- **Better relationships**. If you are talking to someone else, instead of being frustrated by them because they aren't meeting your ideal, you can see them with fresh eyes and notice that they're just trying to be happy, that they have good intentions (even if they're not your intentions), and they are struggling just like you are. This transforms your relationship with the person.
- **Less procrastination**. If you're procrastinating on a big work task, you could look at it with beginner's mind and instead of worrying about how hard the task will be or how you might fail at it . . . you can be curious about what the task will be like. You can notice the details of doing the task, instead of trying to get away from them.
- **Less anxiety**. If you have an upcoming event or meeting that you're anxious about . . . instead of worrying about what might happen, you can open yourself up to being curious about what will happen, let go of your preconceived ideas about the out-come and instead embrace not knowing, embrace being present and finding gratitude in the moment for what you're doing and who you're meeting.[3]

If we lay the lens of these benefits over the task of creating healthy daily post-pandemic routines, it becomes clear that re-entering life with a beginner's mind has the potential of enhancing our overall satisfaction, contentment, and health. The difficulty lies in our willingness to be uncomfortable and vulnerable enough to consider our daily habits non-defensively. This will be an especially difficult task in light of the tenuous condition of our mental health given our exposure to the prolonged distress of the pandemic.

Being self-aware requires immense courage, and making changes to our existing habits commands herculean amounts of grit and resilience. Few are the people who enjoy submitting themselves to the input of others or to the wisdom gleaned from painful life lessons. Adding insult to injury, we've constructed our daily lives in such a way that we can largely avoid self-reflection or feedback and instruction from others. Our devices are, perhaps, the most accessed accomplices in our pursuit to stay distracted, but many other patterns keep us blind to the ways in which our "expertise" at life actually prevents us from thriving.

Adopting a beginner's mind is one way to approach our return to in-person communal living with the health of ourselves and others as a priority. By taking the stance of a novice in our own lives, we can begin to ask ourselves important questions not only about our pre-pandemic habits but also about how we might fashion lifestyles that promote deeper connection with ourselves and those with whom we live in community.

If we don't take the time to thoughtfully examine our pre-pandemic way of life, we become vulnerable to unconsciously idealizing patterns that don't serve or grow us. Just because we found our nightly trips to the bar to be socially satisfying and de-stressing doesn't mean they are the healthiest way to chill out from a long workday or connect with others. This isn't to say that bar nights are bad, per se, just that, unexamined, we may not consider other ways to meet the same needs, thereby diversifying our options and, potentially, de-coupling consistent use of alcohol with stress relief and social opportunity. The examples akin to this one are many. In our rush to re-enter in-person and out-and-about living, it would be easy to become complacent about what is healthy (and what *might be*) given the strong urge to "get back to normal" simply because it's a known commodity (be experts at *what was*).

ASSESSING

To move from mindless "expert" on how to live to being a person starting over with a beginner's mindset, one must make concerted efforts to self-assess. The following questions are a good place to start. Working through these with a childlike curiosity, as opposed to fear or threat, will help illuminate any habits or patterns that have hurt rather than helped.

It's important to prepare ourselves to come to these questions with the mind of a gracious and accepting student. We aren't grading the substance or experience of our pre-pandemic lives. Instead, when we contemplate these questions over a period of days, greeting insights with openness and non-judgment, patterns and themes will emerge to instruct us. This is a time to get our pens and begin taking notes, right here in the margins, and to plan to return often until the patterns make themselves known.

- What was my relationship to my devices pre-pandemic (p-p)?
- How did I care for my body p-p?
- What activities in my p-p life made me feel capable and emotionally stable?
- What activities in my p-p life zapped me of energy or left me feeling incompetent or unhappy?
- What spaces and places did I find myself gravitating to for social connection p-p?
- What spaces and places did I find myself gravitating to for intellectual stimulation p-p?
- Did these spaces offer opportunities to get my social, entertainment, and intellectual needs met in healthy and life-affirming ways?
- How did I soothe myself/tend to my emotions p-p?
- What kinds of balance did I have between time, energy, and effort spent in my embodied life versus my digital life p-p?
- What am I missing most about life p-p? (Be as specific as you can.)
- Were there any costs (personal, emotional, relational, vocational) related to the p-p activities I missed most? (e.g., Family paid a price because of the amount of time I spent at work? Relationships never got deep because I was too busy to invest in them?)

- Adopting a social justice lens, are there ways in which I lived life from a blindness to my privilege? Am I affiliated with systems, entities, or people who act in ways that oppress and marginalize others?

Once we've sat with these questions, and the answers that have floated to our conscious minds, we can begin to determine which parts of "normal" life are worth re-engaging and which parts may benefit from some change. Setting some intentional norms will help if we want to move from the mere habitual living of life to more intention and greater impact over the trajectory of our lives.

SETTING THE STAGE FOR NEW NORMS

Whereas it's easier to establish healthy norms than it is to break bad habits, neither is easy. Onboarding new norms is painful and requires the kind of restraint and intentionality that our pandemic lifestyles have compromised.

We've already covered the difference between habits and norms in chapter 6. Here, let's explore a few actions that will support new norm setting:

1. **Set one new norm at a time**. We can only succeed at so many things at a time. When we feel motivated and inspired, it's easy to bite off more than we can chew. Making sure we onboard only one new norm at a time and letting it be firmly settled in place before beginning another will help us maintain success.

2. **Communicate plans to work on the new norm with those in your daily life**. Ask family, friends, or coworkers for their support and make a plan for how to tap into that support when needed. We are less able to be our patient selves when breaking habits or setting new norms. If we live in context of others, it's only fair to let them know our plans. It's also appropriate to ask for help as long as they are free to say yes or no. Communicating as clearly as possible up front about the changes

we are trying to make and the kinds of input and help we do and do not want is crucial.

3. **Be specific about the overarching norm you'd like to set**. To achieve an intended outcome, it's important to be clear and detailed about what we want and to tease out the behaviors, mindsets, and others that will help us get there. If we've been drinking too much alcohol or watching too much YouTube to get through the days at home, it would be easy to establish the new norm we'd like to set as, "Be more healthy." This kind of goal is too broad to be helpful. In narrowing our focus, we might ask ourselves, "What are the feelings and experiences driving me to drink or binge on YouTube?" When we understand what's beneath our habits, we get good clues as to what norms might help us break them. For example, if we realize through our questions that feelings of restriction (can't leave the house, have limited access to rewarding experiences, and the like) lead us to want to have some kind of sensory experience, we can explore that specific realization and ask, "When do I find myself most likely to feel restricted and turn to drinking or bingeing on YouTube?" If we answer that question, say, with "the evening," then we can be specific about the norms we'd like to set that will offer us healthier alternatives. In the example, this might sound like, "I want to establish the norm of getting outside for fresh air or a walk in the evening, spending the time planning an experience or two that would be pleasurable at night that doesn't include alcohol or YouTube."

4. **Make a realistic plan for how to break your old habit and set the new norm by determining the necessary small steps**. Whereas some people can stop behaviors or ways of being cold turkey, they are the outliers. Habit breaking is difficult and thorny work. We will benefit by researching how to best break our habits and to move with real intention and care in doing so. One surefire way is by breaking down our plan into small, doable steps. We will do best if we don't have to think too hard when we're trying to live in a new way! For this reason, a solid and easily accessible list of those steps and of tools we can access will be a huge support to our process. Using our example above, we might list the following steps to take:

- Every Sunday I will make a list of things that would be helpful to have in my home as alternatives to resorting to alcohol or YouTube for de-stressing. These might be foods and nonalcoholic drinks I love that require no prep, as well as art supplies, games, puzzles, or other things to do with my hands and mind. I'll find a way to keep them at the ready and easily accessible.

- During the week, as soon as I finish work, I will put on my shoes (which I leave by the front door) and walk around the block three times.

- I will leave my phone at home during my walk, leaving space for me to transition from work to leisure time.

- On the last trip around the block, I'll brainstorm some ideas for the evening that would be more meaningful than and equally as enjoyable as drinking or bingeing on YouTube.

- Throughout the evening, if I feel compelled to grab my device to binge or pour myself an alcoholic drink, I'll do a quick scan and ask myself if there is something else with which I could engage my mind and/or body in order to have a sensory experience that doesn't involve one of my usual two options. If touching base with someone would help me with that assessment, I'll call/text (list a couple of names here).

5. **Identify and set up meaningful rewards for keeping up the hard work of establishing the new norm well in place**. Too often we try to grit our teeth and make plans to stop an old way of being and begin a new one in a cold-turkey fashion. As noted earlier, unless we're an outlier, this approach sets us up for failure. We will do much better if we add meaningful rewards and pleasurable experiences to our efforts to onboard new norms in regular, consistent, and pre-planned ways. For example, if we'd like to set the norm of getting up at a consistent time every morning to have some time to meditate before beginning our day, getting everything we need to have a delicious hot drink ready for ourselves upon rising will support our efforts. If we'd like to stop checking our social media in bed before sleep, setting up our night table with a journal,

paper book, or other personally meaningful objects with which to engage will help us feel comforted rather than denied.

6. **Make a loose contingency plan in case you experience a lapse**. This brings us back to our earlier point about gathering support. It's okay to ask for help in maintaining your goals. It's also important to have self-compassion, think flexibly, and avoid all-or-nothing thinking. When trying to make changes, to take a few steps forward and then a few back is normal. If we are trying to eat a healthier array of foods and have a day we eat only Cheetos and Diet Coke, we don't need to panic and shame ourselves. We simply get up the next day and try again. This is the beauty of beginner's mind. Each moment is a new moment in which to begin again.

DOING THE WORK OF NORM SETTING

- **Commit to restarting with a beginner's mind in one area of re-entry**. Identify one part of re-entering the post-pandemic world that you can approach as a learner. Intentionally set aside what you expect and commit to deep curiosity about what this part of life will be like.
- **Work all the way through the establishment of one new norm**. Using the steps laid out in this chapter, sketch out a basic plan for establishing one new norm.
- **Get buy in and set the stage to succeed**. Using the norm you've established above, take steps to ease the path and work to achieve buy-in from anyone who'll be impacted by the new norm. Decide on a start date and commit to it with flexibility in mind.
- **Put some dates on the calendar to review progress**. Set appointments in your calendar for one week, three weeks, and two months out. Use these times to review your progress toward establishing the new norm. Take steps to address anything that isn't working.

8

TECHNOLOGY NORMS
TO CULTIVATE

When boundless new opportunities beckon, figuring out step one can be challenging—especially when we've been under quarantine for as long as the COVID-19 pandemic has lasted. Some clues of where and how to re-engage with life outside our four walls can be found in pre-pandemic research (confirmed by our own guts) about how our habits with technology were impacting our lives. In sum, our near-constant engagement with devices, even prior to the coronavirus's discovery in late 2019, was taking a toll.

Long before lockdown, our attention spans were short and our ability to focus impaired. We passively shied away from spontaneous experiences, choosing curated ones instead. We were experiencing depression, anxiety, and a fear of missing out that was not only correlated with, but also caused by, excessive social media use.[1] The heightened sense of comparison and competition these spaces fueled led us to feel insecure in our embodied interactions, in turn fueling skyrocketing cases of social anxiety across most demographics. Gaming, movies, and streaming offered access to entertainment and distraction 24/7, and our ability to multitask became epic, costing us calm and impairing our emotional regulation skills.

Granted, digital engagement has many positive benefits. We can connect and stay in touch with more people in more places and have access to information and friendships in powerful and expansive ways. We are able to make advances in the world of science and medicine that would be impossible without the internet and the power of modern devices. We are able to offload many of our more manual and menial tasks to make more time for recreation. We can have whatever we want delivered to our door and, basically, have the world at our fingertips.

Once we were confined to our homes because of the pandemic, these benefits became actual lifesavers. We even moved activities to the digital domain that previously were considered impossible to do online: seeing our doctors, going to school, running our businesses, and looking to our devices for 100 percent of our entertainment needs. We quickly lauded the fact that the internet saved the day.

A difficult reality is that, while the internet may have "saved" us during quarantine, our nearly full dependence upon it comes with costs from which we won't easily recover. The more acclimated we have become to offloading our work, learning, communication, and entertainment exclusively to online spaces, the less comfortable we are with being inconvenienced, uncomfortable, and bored in our own skin and in relation to others. This is important because these three states of being—being inconvenienced, uncomfortable, and bored—are largely responsible for our ability to develop grit and resilience, two traits necessary for living a meaningful and not completely self-centered, entitled life. Basically, to be a capable and content human, we must be self-aware, able to foster an accepting relationship with our own selves. Likewise, to be an altruistic person, we also must be able to connect with the humanity of others.

Just because we *can* live in a world where we don't *need* to consider the needs or wishes of others—where we don't *have* to share, or wait, or be still—doesn't mean it's healthy to live exclusively in that world. It just might be that this new opportunity to begin again helps secure us new opportunities to practice being uncomfortable, bored, inconvenienced, and aware of the needs and preferences of others. These experiences can be especially instructive if we take the information gleaned from our time in quarantine and use it to set norms that might put us on a new trajectory toward health. The following areas of our daily lives might be good places to start.

MULTITASKING

In the land of technology, multitasking is aptly referred to as "task switching." The words describe an activity we've passively come to associate with efficiency and skill. In America, and many other places around the world, the ability to multitask is praised, and those who do it well are seemingly rewarded for their efforts. Research studies, however, have exposed multitasking as harmful. In reality, it is a fancy way of saying "distractible" and is associated with dividing attention and behaviors between tasks. This is especially true when applied to the digital domain.

Whether toggling between two (or more) devices at a time or going back and forth between tabs and apps on one device, we digitally multitask our way through much of our days. Overall, a constantly evolving body of research tells us that we almost always take longer to finish tasks and make more errors when switching between tasks than when we tackle one task at a time.[2] In addition, studies show that we inflate our actual ability to multitask, thereby imbuing the behavior with greatness that may not be factual.[3] Stanford researchers wisely state, "Multitasking is almost always a misnomer, as the human mind and brain lack the architecture to perform two or more tasks simultaneously. By architecture, we mean the cognitive and neural building blocks and systems that give rise to mental functioning. We have a hard time multitasking because of the ways that our building blocks of attention and executive control inherently work. To this end, when we attempt to multitask, we are usually switching between one task and another." So, although we aspire to be great multitaskers, the actual behavior may hurt our outcomes and be unworthy of our effort.[4]

Media multitasking isn't just an issue for adults. Given the role devices have played in education during the pandemic, it's important to consider how multitasking has impacted children, youth, and young adults, whose brains are still developing. Research shows that young people spend more time engaging with digital media than adults and that 29 percent of their online time is spent juggling multiple media streams. This leads researchers to report reduced cognitive performance and increased psychological distress in children and youth who digitally multitask.[5] Furthermore, heavy digital multitaskers are showing deficits in working memory;[6] in interference management skills (tasks requiring filtering out distracting information, either from the external environment or the internal world);[7] and in completing tasks requiring sustained focus.[8]

Psychologically, youth who digitally multitask show higher levels of impulsivity and sensation-seeking behavior,[9] higher levels of social anxiety and depression,[10] and lower perceptions of social success[11] than those who report lower levels of digital multitasking. Although some of the studies illuminating these psychological impacts had small sample sizes, anecdotal research supports these findings.

Interestingly, a literature review of many current studies on multitasking found that *perception* is more important for finding success in multitasking than the *actual behavior.*[12] This means that, whereas the *behavior* of attempting to tend to more than one task at a time largely leads to *decreases* in performance, someone who *perceives* themselves as multitasking is more engaged and consequently outperforms those who perceive the same activity as single tasking. To this end, perhaps we could, in taking on one task at a time, consider this multitasking—with the action of doing one thing at a time being one task and the action of learning to avoid distractions the second task.

RESTART UNI-TASKING

- **Experiment with doing one thing at a time**. Set certain periods of a day or a day of the week when you will deliberately do only one thing at a time. If you are washing dishes, you do only that, foregoing listening to music or a podcast.
- **Create physical distance between you and your phone**. Leave your phone in the trunk when you drive and in your pocket or bag when you are out and about. Resist the urge to pull it out at red lights or while waiting in line, taking notice of the moment and what is happening around you.
- **Uni-task with devices**. Use only one device at a time. Close all tabs and apps other than the one in which you are currently working.
- **Turn off notifications**. Turn off all message, email, and news alerts while you work. Set a fixed time of day and amount of time you'll check each.
- **Experiment with mindfulness and/or meditation**. Do a little research and find a mindfulness teacher or system that jives with you. There are many apps and online tools that can give you the basics of mindfulness and offer you short mindfulness meditations. These practices can support single, mindful tasking. It's important, however, to learn a little about how the meditations work rather than simply doing mindless guided meditations with your devices. The goal is to be able to do a one- to three-minute meditation on your own when you need it rather than reaching for your device to guide you.

BOREDOM

Boredom, as an experience, has received a bad rap. We actively avoid it and demonize it regularly. We claim its presence as a cry for help, with "I'm bored" being a call to anyone to offer us anything to save us from being alone with ourselves. In reality, however, boredom is an important teacher we would all benefit from embracing.

Perhaps the most foundational benefit of boredom is that it grounds and steadies us, providing us the opportunity to be deeply in touch with our truest selves. When we can tolerate moments of relative nothingness—those still, quiet spaces between the chatter of thoughts and the doing of deeds—we glimpse our own consciousness and develop the prerequisites of building a strong sense of self. With our "True North," or inner voice, known, we are able to face life's challenges with stability as well as flexibility. A gift in and of itself, boredom also serves as a means to new and healthier ways of being in the world, ways that create the capacity to push through blocks in our thoughts, feelings, and actions. Boredom enables us to be creative problem-solvers.

RESTART WITH BOREDOM

- **Put yourself in the way of boring experiences.** When watching streaming services, resist the urge to "Skip Intro" or "Skip to Next Episode." When an episode ends, wait a full minute before beginning anything else.
- **Prepare to get bored.** Create a boredom corner in your living or work space. Stock it with a few non-cognitive-based fidget or sensory toys, a notepad and pen, and a bottle of water. When you feel yourself resisting boredom, go there for at least five minutes without any devices. Breathe deeply and try to bring yourself fully into the present moment.
- **Commit to ten minutes a day of boredom.** Establish a daily ten-minute boredom practice. As well as you can, let your mind wander and yourself do nothing. If it helps, make sure you bring a notepad and pen with you in case you get stuck with an obstinate thought you can't let go of or an insight you want to capture.

The Dutch have a word, "Niksen," that they define as "doing nothing deliciously." This seems a beautiful way of thinking about boredom. In these months of sheltering in place, however, we've become especially tired of being bored. With nothing but the same environment day after day, we've invested a lot of energy in planning all the ways we will not be bored when the world re-opens. This makes sense. Even still, without forethought and care, we'll find ourselves as overstimulated, overworked, and overwhelmed as when we entered quarantine. Tending to a balanced life in which we intentionally make space for both boredom (pause) and action will give us a leg up on health, stamina, and thriving.

IDLE TIME

Our near-constant scrolling habit, which was a pre-pandemic problem, has only become more ingrained. We finish a meeting with a few minutes to spare and hop over to check the news. We wrap up an assignment before it's due, then hop over to socials to "catch up." A call is shorter than we planned, so we tackle another level of a game on our phones or start several conversations in text. By doing this, we use up all our free moments with mindless scrolling and tech engagement. Five minutes here and ten minutes there add up to hours every week and weeks every year. There is another way.

Becoming conscious or mindful of our habitual reach for devices and apps is a powerful first step toward honoring idle time. Idle time is important because it's time we can use to refresh and restore ourselves if we don't mindlessly fritter it away. It's time rife with potential when we are aware of its presence. This potential, when harnessed and engaged, can be a valuable investment in our emotional and physical well-being. Just as a car is still active when it is stopped but idling, we can be inwardly active when we are awake but not preoccupied with a task or need for engagement. When we create pauses during idle time, stopping our habitual grasp for devices, we increase our sense of personal agency and have greater potential to stay emotionally regulated.

The truth is, we can have a huge payoff when we link (1) our impulse to distract ourselves with online content, and (2) mindful questions such as, "What embodied activity might be more fulfilling than spending this chunk of time scrolling?" "What do I need in the way of self-care in this moment?" or "Would I be refreshed by being present to this moment without distraction?" Let's say we realize we've got five minutes before our next meeting. When we learn to take a pause before grabbing for our phone and going down the digital habit hole, we might use the moment in a more personally powerful way. Perhaps we'd stand up and stretch or unload the dishwasher or step outside for some deep breaths of fresh air. Maybe we'd sit and stare into space, offering ourself a few moments to catch our breath. Regardless, we'd invest a few moments in being intentional about our time rather than simply frittering it away.

When idle time is found outside those small marginal spaces between tasks, we can consider saving it and bundling it together to have a greater chunk of time for a meaningful endeavor. Let's say our Wednesday evenings are always free, and we find that, after dinner, we consistently lounge on the couch surfing the web and social media. If we were to plan ahead, however, we have a chance to feel a real sense of accomplishment by using that chunk of time for something meaningful to our well-being such as working on a hobby, reading a paper book, or going for a walk. Similarly, if we find ourselves getting distracted while working, taking spans of time to engage mindless digital spaces, we can work to reward ourselves for staying focused by adding up that time,

---

RESTART BANKING AND USING IDLE TIME

- **Contain your online time.** Choose a day to preset the times and spaces when you'll interact with devices. Delete all apps and push notifications on your phone and only check media in the predetermined times. Notice the moments of free time when you would have been scrolling mindlessly and tally them at day's end.
- **Put your device down/turn away from the screen every time you finish a task.** When you complete a digital task, make it a habit to set down the device (screen down) or to turn away from a stationary screen. Do a few neck rolls or stretch. Ask yourself to forego instantly tackling your next task.

pushing through with focus, and using the banked time for a deeper digital experience or an embodied one later.

Too often we feel we don't have time for more intentional, slower-paced living. As we wake up to how much time we spend in relatively meaningless and, sometimes, potentially harmful scrolling, we realize how much idle time we actually have. Bundling and banking these periods of time is possible, and using these periods for activities that will improve our satisfaction can add potent purpose and delicious joy to our lives.

BALANCING EMBODIED AND ONLINE SOCIAL ENCOUNTERS AND COMMUNICATION

The rise in global loneliness—being featured in both popular and academic presses even before shelter-in-place orders—is real.[13] Quarantine has only intensified this.[14] Whereas we have found ways of connecting with others online, for many, the loss of the ability to connect in embodied ways has certainly intensified the sense of both loneliness and aloneness.

Our comfort with verbal and in-person communication and encounters is related to a complex mix of temperament, experience, and perceived (and actual) safety of the environment. Some people find great ease in human interaction, whereas others experience anxiety when face-to-face. Certain individuals prefer in-person get-togethers, whereas others would much rather text as their primary form of communication. For many, the rise in digital options for communication has been a major boon; for others, there is an experience of significant loss.

Quarantine has offered all of us, regardless of our interpersonal preferences, a serious decrease in opportunities for shared embodied experiences with anyone outside our bubbles. Although we have certainly benefitted by being able to connect with people online, the preponderance of digital communication has likely significantly impacted our sense of ease and comfort with in-person encounters.

Social anxiety was on the rise around the world before we were quarantined and appears to have been especially prevalent in young adults.[15] With COVID-19 restrictions on interaction, this form of anxiety has hit almost pandemic levels. The dramatic increase in prevalence is widely true for individuals living with clinically diagnosed Social Anxiety Disorder (SAD), but even research subjects who claimed, in an important study, to have no social anxiety scored high enough on the Social Interaction Anxiety Scale to meet criteria for the disorder. Given this, those suffering often don't recognize that their symptoms are features of a treatable condition.[16] This intensifies the problem, as SAD is associated with other mental health issues such as depression, alcohol use disorder, and increased suicide risk.[17]

Although not all people who prefer to avoid social situations are living with a diagnosis of SAD, a large majority of humans facing social situations experience SAD's primary symptoms—including a fear of evaluation and an avoidance of social situations. The former is alive and well in all settings, not the least of which are digital domains, and the latter has certainly been made worse by the fear of social contact during the pandemic. When a majority of people in our surroundings are co-experiencing these realities, there is a cultural cost[18] that comes with the personal pain accompanying this disorder and with social anxiety in general.

Given that the brain "wires together where it fires together,"[19] it is likely we all have been overwriting the regions of our brain involved in moderating our emotional responses to embodied gatherings. Those who consciously or unconsciously feared social evaluation—and avoided social settings as a result—have probably felt relief during quarantine. For all people, however, the simple loss of practice in being with others will likely result in new levels of hesitancy in diving back into opportunities for social contact.

It will behoove us all to recognize that our social skills may be rusty after so much inactivity! We may even feel hesitant to invest the time and energy required for making plans away from home and with others. Being aware of this and beginning to plan will help ensure that we don't fall prey to our new habit of getting all our relational needs met in online spaces. Doing some self-assessment about what type of social encounters were life-giving for us pre-pandemic will help, as will identifying the people with whom it is healthiest to gather. As the social world re-opens to us, invitations and opportunities for in-person get-togethers likely will be plentiful. By thinking ahead, we can make plans to prioritize embodied get-togethers with those with whom engagement is life-affirming. Although the first post-pandemic encounters may be a bit awkward and exhausting, we'll benefit by having thought ahead about spending our relational energy wisely.

RESTART BALANCING EMBODIED ENCOUNTERS AND COMMUNICATION

- **Practice talking.** Consider moving at least some of your communication from text to voice. Make quick phone calls for matters about which you might normally text, simply to regain comfort in spontaneous verbal encounters.
- **Identify ways of calming your social anxiety that you can practice as you re-enter social settings.** Practice some quick and effective self-soothing techniques so you have them ready when you re-enter the potentially awkward reality of in-person experiences. (See the Self-Soothing section later in this chapter.) Simple breathing techniques will be especially helpful. Square breathing is easy to learn. To practice this technique, inhale for the count of four, then hold your breath for the count of four, then exhale for the count of four, then hold your breath again for four seconds, then repeat the breathing pattern—all while picturing the breaths and holds as the four sides of a square.

SOLITUDE VERSUS HYPER-ESCAPISM AND INTERNAL VERSUS EXTERNAL LOCUS OF CONTROL

Very likely, most of us feel we've had far too much solitude during the pandemic. There's nothing like going from zero to one hundred in a split second, which is exactly what we did socially when shelter-at-home orders were announced. Unfamiliar with spending large quantities of time with ourselves, we suddenly realized how much distraction our out-and-about lives offered. Given that, even when the whole embodied world was available to us, we still split our attention between it and our devices; the pandemic push to socially distance merely amplified our tendency to be overstimulated and escape into digital spaces.

Solitude doesn't come naturally for many of us, but the ability to be with one's self fully and without distraction is a skill well worth developing. The set of abilities that comes with this way of being is profound. When we have a developed self-knowing awareness—being honest with ourselves about our strengths and weaknesses—we are much more capable of weathering the stressors and distractions in life. Unfortunately, our reliance on our screens to get us through the day eats away at our ability to live in this optimal way.

When we are self-aware, we live from what psychologists refer to as an internal locus of control. This means that we can tune out the external world when we need to in order to find internal balance, comfort, and wisdom. People who rely on this kind of grounded or centered way of living are capable of knowing their own strengths and weaknesses as well as knowing how to work with their thoughts and feelings in healthy ways.

Life as we know it does almost everything possible to keep us from living from our internal locus of control—and our devices exacerbate this reality. When we are sad, we distract ourselves with music or movies. When we face conflict, we attempt to text our way out of it rather than suffering through the awkwardness of a difficult conversation. When we're bored, we can entertain ourselves with a constant stream of videos or games. The internet provides us twenty-four-hour access to digital forms of information, connection (which turns to comparison), and entertainment that rescues us from having to get still and quiet and look inside for what we need.

This makes for a fragile life. If we are reliant upon the distraction or affirmation found in "likes" and "follows," we must work tirelessly to earn them—and we've been doing plenty of this during the pandemic. This leaves us bereft of the skills we need to affirm and comfort ourselves, so we look to digital spaces instead. If we feel shaky when we leave our phones at home or can't access what's happening in our MMORPG (Massively Multiplayer Online Role-Playing Game) world, we passively confirm that our digital lives are more compelling than our embodied ones. When the number of texts or DMs (Direct Messages) we receive in a day determine, even unconsciously, our desirability in the world, we are reliant upon everyone outside ourselves for our well-being. This puts us completely out of control of ourselves—and the random results of our algorithms in control of our very well-being.

The alternative is to live from a solid sense of self, which can be developed only by offering oneself opportunities for quiet self-reflection and truth-telling. This translates into living from an internal locus of control whereby we develop the certainty that, far from being perfect, we are at least capable. We know we can count on ourselves; we can balance a sense of our strengths with our awareness of our weaknesses. From this still center at our core, we remember we are capable of figuring things out, capable of comforting and soothing ourselves, and capable of finding ways of stimulating, educating, entertaining, and developing ourselves. We learn we can depend on ourselves even more than we can depend on our online connections.

We've had months and months to live from a uniquely heightened, pandemic-inspired external locus of control. Living online, while convenient, has severely impacted our mental health and sense of agency. It's also exhausted us. While we've had more time with ourselves than ever, we've likely also distanced ourselves from some of our bigger feelings and thoughts in an attempt to keep going in such distressing and difficult times.

As we anticipate a return to greater freedoms, it would be wise for us to make a plan for how we might maintain a modicum of our time and energy for self-reflection, self-engagement, and self-improvement as we rush to re-engage with the wider world. We've expended a lot of our energy simply getting through this crisis. Rather than invest it all externally from here out, let's make sure we honor and privilege the fact that we've experienced personal and communal trauma that won't just go away when the world re-opens.

EMBODIED SPONTANEITY

Having the freedom to be spontaneous keeps us flexible. Our reliance upon our devices to curate and plan our experiences—along with the way algorithms determine the kinds of information and opportunities we are fed—had begun robbing us of spontaneity long before the COVID-19 outbreak. Rather than stopping at restaurants that looked

---

RESTART WITH SOLITUDE AND AN
INTERNAL LOCUS OF CONTROL

- **Talk with a professional or a wise friend who can help you think through the ways you've coped with distancing and your sense of self.** It's common to hear people say now, "This has been hard on me, but I had it easy compared to [fill in the blank with people who had even more challenges than you did in getting through quarantine]." While this may be true, each person's challenges are very real and have had an impact.
- **Take a time-out before diving into social media.** Each time you consider logging on to social media, set a timer for two minutes instead. In that two minutes, sit quietly and identify two things you value or appreciate about yourself.
- **Curate your feed.** When you engage social media, pay close attention to how each profile or person you follow or encounter makes you feel, unfollowing or silencing accounts that hurt you. Begin to curate your feed in a way that inspires more expansive thought, self-compassion, and ease.
- **Get comfortable with solitude and silence.** Practice tolerating solitude and silence. If finding solitude or silence in your daily life is difficult, do a bit of research to find quiet spaces and times. Outside in nature may be one of these places for you. Also consider churches, synagogues, mosques, or meditation centers. Plan at least one day per month when you can trek to one of these locations to practice tending to your inner world.

---

and smelled great, we Yelped our way to the perfect establishment, deciding what to order long before we arrived. Hesitant to "waste our time" on books or movies we might not love, we trusted recommendations built by our algorithms or read umpteen reviews before embarking on our next read or watch. Over and over again, we went for the perceived forms of "certainty" rather than the unknown. This has only become a greater habit over the course of the pandemic.

Our proficiency with being flexible and resilient would be enhanced if, every once in a while, we risked less-than-perfect food, books, and movies by trusting our guts in the moment. Knowing all we want to know prior to embarking on experiences takes away the opportunity to roll with what is—which, in turn, helps us build grit. When we can tolerate a bit of time spent trying a new book genre in order to broaden ourselves, we grow our ability to try new things and to trust our own gut more than the reviews of others.

Years back, I noticed a "Reflexology" sign in a neighborhood I was driving through. Hungry for a completely spontaneous experience, I asked a friend to research if the spa was safe and what I should know before stopping by. When I heard back that all was well and safe and that I should wear loose pants and a T-shirt, I made an appointment and had an incredible spontaneous experience that took me out of my normal routine and nourished my sense of adventure. We all benefit from these kinds of safe spontaneous experiences.

To challenge ourselves to try some things like this to refresh our novelty-starved, post-pandemic beings, we might make a list of things we thought we'd never do. This could include things such as skate at a roller rink, try an obscure international cuisine, pick out a book solely based on its front cover, make a meal out of only what we have in the fridge, take a dance lesson, race a remote-control car on a track, have our palm read, or any other experience we might consider "edgy." By finding one or two things to try, we can encourage great growth. Committing to not thinking too hard, instead, making a basic plan to do the things, setting dates, and following through. If the thing you'd really like to try requires some information about safety or preparation, consider asking a friend to do the research for you, assuring it is safe and reliable, and telling you the minimum of what you need to know.

RESTART WITH SPONTANEITY

- **Explore a new place.** When it is safe to do so, identify a new part of town to explore by foot or choose a place to visit or experience to have by "feel" (not research).
- **Dip your toe in the water of spontaneity.** Make the spontaneity list suggested above and commit to dates to trying one or two of the activities by.

As our ability to re-engage with embodied offerings expands, it would be wise to reconsider our relationships with spontaneity versus curation and habit. Finding and committing to moments when we can expose ourselves to a diverse set of experiences, trusting we will have (or develop) what we need to handle those experiences, may be key to developing the traits of resilience, grit, and flexibility—qualities that enhance our sense of satisfaction with ourselves and our lives.

SELF-SOOTHING

Too often, we substitute stimulation for soothing. We feel overwhelmed and exhausted, so we turn to our preferred streaming service to entertain us. We feel sad or unsettled, so we lose ourselves in an immersive video game. We feel bored (and, therefore, uncomfortable), so we scroll, far longer than we intended, through social media. The problem is that none of these actions actually address our original emotions.

Aside from platforms and devices made specifically to calm our nervous systems, technology almost always stimulates us. Digital activities we consider "relaxing" or soothing are, most often, just distracting us. We turn our attention outside ourselves and away from our feelings and allow our devices to capture our attention, thus dulling the emotional experience we wish to avoid. Over time, this can grow into the harmful habit of stuffing or denying our feelings, which can lead to greater distress and overwhelm.

A healthier alternative would be to put a literal pause between our impulse to grab our phones, remote controls, or game controllers. When we do this, we have space to quickly touch base with our internal experience and identify what we're feeling. A great tool for doing this is the HALT technique originating in twelve-step programs. In a pause, we can ask ourselves if we are *H*ungry, *A*ngry, *L*onely, or *T*ired (or any other number of feeling states). If we answer yes to any of these, we can then take a few moments to actually address our feelings before distracting ourselves with technology. The result will be a growing comfort in naming our feelings and taking actions that will actually address the feelings. Over time, this will help us become increasingly capable of dealing with our emotions head-on rather than simply trying to distract ourselves from them.

RESTART WITH SELF-SOOTHING

- **Identify your healthy grown-person "pacifier."** Think back to childhood and what you remember (or have heard) about what calmed you when you were upset. Was it a bath or a blanket? A book or being held? As well as you can, identify ways you can approximate those experiences in your current space. When you feel big feelings, give these a whirl before pulling out your device to distract you.
- **Learn and practice breathing well.** Learn some breathing techniques and practice them.
- **Tend to all of your senses.** Consider each of the senses and find ways of addressing each in calming ways. Be specific and gather what you need to have on hand to easily access these soothers.

9

POST-PANDEMIC LIFE NORMS
TO CULTIVATE

"We plan, God laughs" says the old Yiddish proverb. We would all be wise to hold this sentiment close as we embark on our post-pandemic lives. No matter how much we plan, we're going to face constant change in the months and, possibly, even years ahead. There are, however, some things that COVID-19 has taught us and some actions we can take to help us do this process of restarting well.

SLOW DOWN

One common experience among a majority of us during quarantine was an appreciation for a forced slower pace and defensible reasons for saying "no." Even for those who enjoy being on the go, our pre-pandemic pace felt exhausting and unsustainable. We were always moving, always going, always multitasking, and never quite catching up. When shelter-in-place orders made slowing down mandatory, many of us were relieved. While it was difficult and scary, the forced shutdown offered a surprising opportunity to live at a new pace.

For those who craved a slower-moving life prior to COVID-19, or whose temperaments bristle in a moving-at-the-speed-of-light pace, it was a huge comfort not to have to try so hard just to fit in. This group has gotten a much-needed reprieve from having to push themselves to be people they aren't. For them, this time of re-entry will likely be peppered with anxiety about how they will begin saying "no" to social offerings that they (understandably) lied about missing during lockdown. On the other hand, for those who prefer a breakneck pace, there will be an urge to re-enter with wild abandon, ignoring any internal calls for pacing.

Regardless of the appreciation for, or discomfort with, the slower reality that shutdown brought, most of us settled into new rhythms during quarantine. We played more board games, watched more shows, engaged more home-based hobbies, went for walks, and took naps. Even among the huge stress, this time has provided a sense of personal restoration for many.

One of the biggest rewards of moving at a slow pace is that it connects us, often by default, to what is most important. When we are paying attention to life, we can make more intentional choices about tending to what we care most about. When we are keeping pace with a rhythm that doesn't allow time for introspection and intentional decision-making, we are more likely to live mindlessly, tending to that which is in front of us more than that which is important. Sometimes we even find ourselves in the "go fast and break things" mentality driving so much of the world's emphasis on constant productivity. This time in history is the perfect time to evaluate whether we want to align ourselves with the value of speed or that of intention.

MAINTAIN A NEW, SLOWER PACE

- **Assess how it feels to rush and determine if it's necessary or self-imposed.** Think back and identify a time you've done a task in a hurry to meet a deadline or self-imposed goal. How did the time crunch impact the quality of the "product"? Was the rush self-imposed by procrastinating or other reasons in and of yourself? Then identify a time when you took a bit more time. How did this process feel? Did you experience a different level of contentment or quality as a result of the extra time?
- **Commit to a weekly hour of doing only one thing at a time.** Given that our central nervous systems benefit from breaks from our fast-paced, always "on" way of being in the world, set a timer for once a week to do either one (non-screen-based) thing at a time or absolutely nothing at all. As you can, integrate shorter (or longer) periods of doing this throughout the week.

BE BOLD IN LOVING WHOM AND WHAT YOU LOVE
AND WASTE NO TIME IN DOING SO

If nothing else, COVID-19 brought our mortality into stark relief. When it hit, it didn't offer people a chance to work through regrets or grievances. We weren't warned to cherish the last meal in a restaurant or good in-person conversation we experienced before sheltering and distancing for months on end. We didn't know we'd go a year without hugging. We didn't know we'd lose loved ones without getting to have an in-person goodbye. As a result, we've come face-to-face with our own limits and mortality as well as that of those we love.

For many of us, the virus and resulting safety protocols revealed our "inner circles" and clarified our closest friends more than ever. We've likely lost some relationships while others have deepened and blossomed. This has been painful and beautiful in equal measures. We come to this moment feeling let down by people we were sure we could trust and were surprised by how people previously at the fringes of our inner circles moved deeply inside of them. Regardless, it's likely that we are facing re-entry with a different depth to our awareness of whom we love. It's important to lean into this.

As the world re-enters, it will be easy to feel pressure to accept invitations for reconnection from people we felt distant from during quarantine. It'll be important to stop and take stock before accepting. If insight makes it clear that a relationship wasn't a healthy one over the course of the last year, spend some time really evaluating if this is the case. If, after evaluation, you determine that it is best to set boundaries in this relationship or discontinue it altogether, think through what is required of you. In some cases, you can simply decline reconnection and, in others, you may need to be more clear. This could mean writing a note that says something like, "Thank you for reaching out. I'm re-entering post-pandemic life with great intention and feel a need for space. I'm grateful for our season of connection and wish you the best."

We only get one shot at life and we need to use it wisely. It's time to get busy doing this. It's time to work through conflicts or disillusionment with those who are safe to do this with and to re-align our relationships with our values. It's time to show up in meaningful ways to those who need us, including ourselves. It's time to take our commitments to those we are attached to seriously, making as much time and space for them as we do for our more self-centric pursuits. When we say yes to something or someone, it often means we'll need to say no to something or someone else. We need to be intentional about how we do this and we need to be part of the equation. If we aren't saying yes to grounding our love of others in care and love of ourselves, we will not succeed anywhere.

It's admirable to want to love and serve all people equally. The reality is, however, that we have only so much time and energy to offer and at least some of this needs to go toward nurturing ourselves. When we push past this, ignoring the reality of our limits, we risk burnout as well as the eventual resentment of those we serve.

To align yourself with reality, make a list of all the people you currently interact with or are connected to. When your list is as exhaustive as you can make it, draw three concentric circles and go back though and identify which people are part of your inner circle, which are part of the next ring out, and, finally, those who exist in the furthest out "acquaintance" ring. Write the names in each ring.

Looking at the circles, see if there are more people in the center ring than you can realistically care for/nurture relationships with. Are there any that you can actually nudge closer to the acquaintance ring? Are there any people in the closest ring that are not healthy relationships for you or that zap you of energy and life in unhealthy ways? What would it take to re-orient your relationships with them to look more like those in outer rings? Keep working at this until you have a manageable number of relationships in the closest ring. If you need more people in that center ring, look at the outer circles to look for people with whom you might be able to nurture deeper connection. Set times on your calendar to come back to assessing how it's going re-aligning your relational life with the realities of your limits and needs in mind.

This prioritization of grabbing hold of that which is most important to us includes activities/actions as well as people. COVID-19 taught us, in many ways, the importance of meaningful experiences. Mostly, it taught us not to put them off. If there are things we've been waiting to experience, learn, or try, this time of re-entry may be the perfect time to do so. We know, to our cores, what it feels like to *not* have the freedom to engage experiences that would give us life. Let's not waste time getting to them now.

Making time to nurture our own beings in this time of immense change involves making hard choices, committing to taking action in our relationships and lives and doing that in intentional ways. If we don't do this, we'll weaken our ability to be present to others as well as to the tasks we will be required to undertake. Taking steps to make sure that that we prioritize some time for ourselves, those in our innermost circle, and the activities that we know will feed us as we move back into a more engaged life must be a priority in this time.

BEING BOLD IN LOVING AND BEING

- **Make the culling of your community "official."** Determine friendships that may need to be let go of and what your part is in creating an end point for them.
- **Commit to cultivating a loving relationship with yourself.** A commitment to caring for yourself and building your authentic self up is the foundation of loving others well. You may feel tired of time in your own head after months of isolation, but prioritizing time like this for the long run will help you be able to be grounded and manage stress during re-entry and beyond.
- **Prioritize in order to help you know how to say yes and no.** Become conscious of how your yes's translate into unconscious no's. Be intentional about this exchange.
- **Tend to habits of codependency.** Codependency is the excessive emotional or psychological reliance on a person, to whom we tie our own well-being. If you find yourself giving inordinate amounts of time and energy to someone, hoping that their well-being will assure yours, it may be an indication that the relationship is a codependent one. Seek help to detangle yourself from this pattern.

KEEP ALWAYS BECOMING A BETTER COMMUNICATOR

One of the best tools we have for maintaining a healthy relationship with ourself as well as with others is the ability to communicate in honest and respect-filled ways. COVID-19 hit at a time of immense global and national unrest that was actively splitting many communities and families. As a result, interpersonal communication was, too often, a source of great distress. Quarantine made casual conversations less available and topics that had normally been danced around became fodder for full frontal attack. It was hard to find inroads with many people who were doggedly certain about letting the world know that they had nothing to learn from anyone who disagreed with them. Broken bonds, horribly hateful language, and the condoning of gaslighting and lies took a significant hit on the reliability of respect-driven conversations.

The pandemic revealed gaping holes in our collective ability to handle conflict well. The violence of much of the communication that took place during this time was palpable. As we move into new shared spaces, it will be important to find ways of embracing the conflicts that are natural to life and working through them in ways that will build community rather than tear it down.

Nonviolent communication (NVC) offers a framework and tools that can help to this end. Developed by Marshall Rosenberg, NVC is defined as: "a 'language of life' that helps us to transform old patterns of defensiveness and aggressiveness into compassion and empathy and to improve the quality of all of our relationships." It involves engaging four components (Observation, Feeling, Needs, and Request) and two parts (Empathy and Honesty) to enrich our conversations and help us resolve conflicts. Rosenberg's basic "script" goes like this:

When I see that_____, I feel _____ because my need for _____ is/is not met. Would you be willing to _____?

We would all do well to learn more about NVC. Until that time is available, we might, at the very least, begin practicing this script for use when we need it most. This would go a long way toward helping us work healthily with conflict and also deepen your connection with others.[1]

An example of a completed script might read something like this: "When I see that you are holding so steadfastly to your opinion, I feel threatened because my need for being heard is not met. Would you be willing to repeat back to me what you are hearing me say so that I can understand if I've been clear?" or "When I see that your body is communicating strong anger, I feel afraid because my needs for safety aren't met. Would you be willing to come back to this conversation after you've released some of your anger elsewhere?"

In NVC it isn't this script, per se, that's important. It simply helps you make non-judgmental observations (the first line), express your feelings (second line) and needs (third line), and make a request in a nonviolent or defensive way (fourth line). As you practice you will find ways of expressing these four important things in your own ways.

Being a good communicator isn't something we are necessarily born with. It is something, just like resilience, that we all have the capacity for but must work to develop. This means paying attention and learning, in formal and informal ways, about what we do well and what we might need to get better at. By doing a self-assessment and/or asking those who are in our closest proximity to help us identify the ways in which we do and don't listen and discuss/express effectively, we can identify the things we need to work on. From there we can embark on some education and practice. Since communication is the tool we'll use to rebuild our communities, it's important that it's in good, reliable shape.

We have an incredible opportunity to engage the lessons we've learned from this time. Working together to become both better listeners and more effective speakers, we can begin the work of healing ourselves and our communities.

MAKE (AND KEEP) SELF-CARE ROUTINES

The words "self" and "care," taken together, usually provoke one of a few reactions: eye rolls, exclamations of "Oh Brother," visions of middle-aged, white women day-drinking and getting facials, and privileged folks laying around a pool. These are not the kinds of self-care I'm referring to here. What I am wanting to encourage is the kind of intention that Audre Lorde had when approaching self-care. "Caring for myself is not self-indulgence," she said, "it is self-preservation, and that is an act of political warfare."[2]

If we don't care for ourselves, we can't hope to love others or serve our global neighbors well. This is a primary reason why quarantine was so hard; we lacked time, space, and energy to care for ourselves in meaningful ways at the exact time when we were caring for others in

RESTART AS AN EFFECTIVE COMMUNICATOR

- **Beef up your communication skills**. There are many webinars and online/in-person courses you can take to improve your communication skills. The goal is to learn *both* how to be a better listener *and* speaker and to be flexible in both areas. Often, small tweaks in how we approach either task can make a huge difference.
- **Take responsibility**. Use "I" statements as much as possible. Instead of saying "You make me so angry!" say "I feel so angry in response to that." Instead of saying, "Those jerks. They totally ruined my day," say, "I am having a really hard day. I need to find a way of shaking off that hard interaction." This helps us take responsibility for what we can control and let go of what we can't rather than keeping us in victim mode.
- **Work for dialogue rather than debate**. In a dialogue the goal is for people with differing opinions and outlooks to come to a shared understanding, even if that understanding is to disagree. Dialogues help deepen every participant's understanding of the topic at hand. The goal of debate is for one person or viewpoint to win. Work actively to engage and respond to others from a position of being in a dialogue.
- **Use the "Ouch/Oops" technique**. Coming out of the Diversity, Equity, and Inclusion movement, this technique offers a way of interrupting the strong emotions inherent in confronting microaggressions (a statement or action that is indirectly, unintentionally, or subtly discriminatory). This works best when people in a relationship or a defined group agree to say "Ouch" when a microaggression is made. This instigates a moment of pause, and the person who has made the offending statement can respond with "Oops" and try their statement again. It offers a shared way of communicating in an emotionally charged situation that could otherwise derail the conversation.
- **Find a way to engage and be comfortable with conflict**. Revisiting the information on NVC, write out or say out loud a few versions of Rosenberg's script using real-life examples. The blank script looks like this:

    When I see that_____, I feel _____ because my need for _____ is/is not met. Would you be willing to _____?

- **Watch your self-talk**. The way we talk to ourselves has a lot to do with the way in which we move through time and space. If we talk down to ourselves or speak to ourselves in inflated ways that are out of touch with reality, we'll be likely to be content with being dismissed or idealized by others, respectively. This will impact the way in which we do or do not form healthy attachments and is worth real introspection.

big ways. Needs for distancing, mask-wearing, washing our hands, and foregoing social gatherings were ways we faithfully cared for others, and all of these things came with a cost to the energy and time we had to tend to ourselves. True self-care often requires solitude and independence, planning and access to certain experiences, all things in short supply throughout the pandemic. As a result, we let it go, and we are all paying a collective price. We're exhausted, teetering on (or fully fallen into) burnout, and bereft of the energy that self-care requires.

Self-care is, perhaps, best rooted in the idea that we can only be healthily present to our relationships, work, and lives when we have first been present to ourselves. It works on the "put on your own oxygen mask before helping other with theirs" model. Self-care that sustains rather than entertains isn't about big splurges but is, instead, something we can best achieve systematically and proactively in small, manageable ways. Waiting until we're overwhelmed or burned out to tend to self-care only makes it harder.

Given the hypercompetitive, success-driven, and curated lives we lead, many of us have historically privileged striving and grinding. We are on the treadmill and the ladder, trying to keep up or get ahead in a world that is rife with competition. Slowing down or making space to get grounded feels too costly to prioritize and, in the time of quarantine, hasn't even felt like an option.

My friend Brenda practices self-care by pouring her Coke into a cut-crystal goblet, grabbing a Mr. Goodbar candy bar, and taking herself to a quiet corner in her house to enjoy both. Another friend goes out to the garden to dig his hands into the earth. Hot baths, a few minutes of deep breathing, and letting a calming song sink deeply into us are also relatively accessible options. The goal doesn't need to be a spa day. Instead, we need to find small ways of redirecting our attention away from the needs of the world and toward our own well-being in consistent and easily available ways.

The difficulty is that post-pandemic self-care habits will emerge if we aren't intentional about setting norms. This means that now is the time to evaluate our unconscious ideas about self-care and to challenge

---

RESTART WITH SELF-CARE

- **Examine the roots of your thoughts about self-care**. Do you have negative associations to the words or positive ones? Is self-care related to weakness or incapabilities in your mind? In what ways did your parents or important others think about/value/devalue self-care? Look these over to understand and uncover any potential biases to the idea of making intentional time for self-care.

- **Define your why**. If there's no real reason to care for yourself, you are less likely to do it. Take some time to make a list of reasons why it is important to take good care of yourself. Make sure that your entire list isn't based on others (e.g., "so that I can take better care of my kids").

- **Become conscious of your internal message indicators**. We are all highly attuned to the alarms and notifications that our devices deliver to us. We are, however, less aware of the message indicators that emanate from our own beings. Tuning in to our bodies, minds, and hearts more intentionally will help us know how to identify what we need in the way of self-care.

- **Make some lists**. On a piece of paper, make three columns. Label one "Body," one "Mind," and the final one "Heart." In each of these three categories, list self-care actions that you could take to address that particular part of your being.

- **Set the stage**. Self-care practices are not easy to maintain. Gathering any supplies you might need, creating spaces in which you can do your practice, and making appointments on your calendar will all help you maintain your practice.

- **Think/plan ahead and/or ask for accountability**. Take a look at your calendar for the next month. Identify any experiences or commitments that will be especially depleting and make some plans, before and after, to offer yourself some grace and peace. Schedule time specifically for this and make any appointments that you might need to help.

them. This new beginning is a wonderful opportunity to determine what has and hasn't worked in the past to keep us healthy and thriving—body, mind, and soul—and to commit to making an intentional self-care plan that will be sustainable.

Noted Scotsman Ian Maclaren wisely said, "Be kind to everyone, for theirs is a difficult journey." As we all re-enter the wide world after months of quarantine, this could not be a more fitting mantra.

The weeks and months ahead will likely be filled with all manner of stress and heartache, fits and starts. We are far from over the finish line in this marathon, and hundreds of thousands of people still live with lasting COVID-19 symptoms. Millions more live with the realities of COVID-19–related loss. New strains will emerge and people will continue to suffer . . . *and* (not *but, and*) . . . we will be offered myriad opportunities to restart and begin again. We'll find ourselves facing exciting and terrifying opportunities right up against the mundane and benign. We'll mask up for the one billionth time, wishing it were all over, and finding tiny glimmers of gratitude for things we've learned that could have been taught only in this time.

In many ways, we've all been forged in a crucible, which is defined as a situation of intense trial, in which different elements interact to bring about the creation of something entirely new. This new something can emerge only as we go, leading us to feel the uncomfortable reality of walking headlong into the unknown. At least we'll be doing it together.

Beginning with ourselves, then those to whom we are attached, and, finally, reaching out to the world around us, we'll benefit from thoughtful introspection and wise norm setting as we restart. We'll do better by privileging healthy amounts of time for planning, intention setting, and reviewing the success or failure of our efforts. We'll also do better by leading with ample amounts of grace, flexibility, and care.

Ours is a difficult, exciting, overwhelming, relief-filled, excruciating, grief-tinged journey . . . may we embark on it with the greatest of care.

RESOURCES

HOTLINES

SAMHSA (Substance Abuse and Mental Health Services
Administration) Mental Health Hotline
800-662-4357

National Suicide Prevention Hotline
800-273-8255

National Domestic Violence Helpline
(https://www.thehotline.org/)
800-799-7233

Partners HealthCare COVID-19 Hotline
617-724-7000

SAMHSA Disaster Distress Helpline
800-985-5990

SafeLink: 24/7 Crisis Hotline
877-785-2020

TEXTLINES

Crisis Text Line
Text "HOME" to 741741

FREE APPS

COVID Coach App (National Center for PTSD)
https://www.ptsd.va.gov/appvid/mobile/COVID_coach_app.asp

Mindfulness App (UCLA Mindful Awareness Research Center)
https://www.uclahealth.org/marc/ucla-mindful-app

WEBSITES

For Help Finding a Therapist:
https://www.samhsa.gov/find-treatment
https://locator.apa.org
https://istss.org/public-resources/find-a-clinician.aspx
https://www.psychologytoday.com/us/therapists
https://www.aacap.org//AACAP/Families_and_Youth/Resources
    /CAP_Finder.aspx

Institute for Mindfulness-Based Stress Reduction Approaches
(find a practitioner, class, or just read about this very helpful set of tools
and techniques): https://www.institute-for-mindfulness.org/offer/mbsr

The International Society for Traumatic Stress Studies
(wonderful demographic-specific resources for employers, parents,
teachers, health-care workers, and individuals about the complexities of
restarting): https://istss.org/public-resources/covid-19-resources

Nonviolent Communication Self-Guide
https://www.cnvc.org/online-learning/nvc-instruction-guide/nvc
    -instruction-guide
To find an NVC practice group in your area: https://www.cnvc.org
    /trainings/practice-groups

Medical University of South Carolina
(fact and tip sheet for dealing with COVID-19–related anxiety):
https://istss.org/ISTSS_Main/media/Documents/Covid.pdf

COVID Survivor Groups
COVID Survivors for Change: https://CovidSurvivorsForChange.org
Body Politic: https://WeAreBodyPolitic.com

Mutual Aid
Pandemic of Love: https://www.PandemicOfLove.com

ACKNOWLEDGMENTS

Since publishing my first book, I have read the author's gratitude statement in every single book I've read. I am more aware than ever how many people it takes to get a book into the world, and I am humbled by this.

Thank you to my husband, Thomas, without whose fierce belief and investment in me, I never would have taken the kinds of risks that would find me in this place that allows me to write. You think more of me than I do, and this humbles and sustains me. Without you, there is no us, which would be a bleak reality in which to live.

To my children, who have taught me more than anyone else on this earth, I am deeply grateful. Your feedback, lessons, and encouragement have made me who I am. The fact that you three (and, actually, more) also read early drafts of this work and told me to keep going is an unbelievable gift. I have become me in your midst and with your love. With you in my heart it is never "raining."

To my editor, Ruthie, who tells me the truth and believes in me and my words even when I fall prey to self-doubt . . . you are a beam of light that radiates everything we all need in this world. Thank you for sharing your light with me and sprinkling your sunshine all over my words.

To my friend Kim, who lost both of her parents to COVID, and to the participants in the weekly COVID Survivors for Change groups that have informed me and broken my heart, I bow deeply to honor your losses and thank you for trusting me with your stories and profound grief. Jennifer, Esmeralda, Angela, Marjorie, Ed, Shelley, Susan, Consuelo, Kim, Chanel, Kevin, Kelly, Shannon, Marlene, Kpana, Amanda, Andrea, Monika, Debra, Kristin, Ian, and all the rest of you . . . my heart is knit to yours! Chris and Amy, thank you for inviting me into this sacred circle, and Brenda, Laura Lee, and Brandon, thank you for invading it in beautiful ways with me.

To Representative Lisa Reynolds, my physician/legislator friend who joined me in offering early active support to our community throughout the pandemic and read early manuscripts, and to Cassie, my nurse love, who trusted me with her emotions and trauma while serving on the front lines, thank you for teaching me and for helping me find my voice on this topic.

To Jennifer, who speaks my soul language and encouraged me to write the book that I *needed* to write, 143! To Jen, who offered me the space that I needed to write said book in a time crunch, thank *you* for your unbelievable friendship. Without you two Jens, this book would not exist. To Vana, who, quite literally, saves me every single week, I don't have enough words to thank you. To Candyce, Amy, Thomas, Bruce, Lisa, and Cory, who offered feedback and technical help, thank you! Huge gratitude to Tiffany, Brenda, Emily, Debbie, Joshua, and Jackie, who offered self-care reinforcement when I couldn't.

To Suzanne and the team at Rowman & Littlefield, thank you for taking a chance on me and this wild idea. To Emily, Deborah, and Susan in advance, thank you for all that you'll do to help me get this out into the world. I'm really, really grateful.

To Tamela Gordon, whose hour with me on Zoom in December of 2020 reminded me of every single thing that was important and possible, I am gobsmacked by the way in which you offered yourself as a cup, serving the elixir of the Divine's message and call. Without you and your gift of intuition and truth speaking, I would never have doubled down or followed the niggling need to write this book. I will be indebted to you forever . . . as will every person helped by this book.

NOTES

INTRODUCTION

1. https://complicatedgrief.columbia.edu/professionals/complicated-grief-professionals/diagnosis/.

2. https://www.ncbi.nlm.nih.gov/pmc/articles/PMC3384440/.

CHAPTER 1

1. John, Ratey, "Welcome to John Ratey M.D. Cambridge, MA," accessed March 8, 2021, http://www.johnratey.com/.

2. "What Is Privilege," YMCA of Greater Cleveland, Ohio, March 4, 2019, https://www.ywcaofcleveland.org/blog/2019/03/04/what-is-privilege/.

3. "Health Equity Considerations and Racial and Ethnic Minority Groups," CDC.gov, updated April 19, 2021, https://www.cdc.gov/coronavirus/2019-ncov/community/health-equity/race-ethnicity.html.

4. Sarah True et al., "Overlooked and Undercounted: The Growing Impact of COVID-19 on Assisted Living Facilities," September 1, 2020, https://www.kff.org/coronavirus-covid-19/issue-brief/overlooked-and-undercounted-the-growing-impact-of-covid-19-on-assisted-living-facilities/.

5. SAMHSA, "Intimate Partner Violence and Child Abuse Considerations during COVID-19," February 27, 2021, https://www.samhsa.gov/sites/default/files/social-distancing-domestic-violence.pdf.

6. Michele W. Berger, "Social Media Use Increases Depression and Loneliness," Penn Today, University of Pennsylvania, November 9, 2018, https://penntoday.upenn.edu/news/social-media-use-increases-depression-and-loneliness.

7. Doreen Dodgen-Magee, "Grumpy, Sad, and Anxious: Week 9 of Quarantine," *Psychology Today* (blog), May 13, 2020, https://www.psychologytoday.com/us/blog/deviced/202005/grumpy-sad-and-anxious-week-9-quarantine.

CHAPTER 2

1. Chimamanda Ngozi Adichie, "The Danger of a Single Story," TED Ideas Worth Spreading, July 2009, video, 18:33, https://www.ted.com/talks/chimamanda_ngozi_adichie_the_danger_of_a_single_story.

2. "Social deprivation," APA Dictionary of Psychology, American Psychological Association, accessed March 8, 2021, https://dictionary.apa.org/social-deprivation.

3. Nicole Karlis, "Is the Pandemic Making Our Social Skills Decay? Psychologists Think So," *Salon*, February 5, 2021, https://www.salon.com/2021/02/04/is-the-pandemic-making-our-social-skills-decay-psychologists-think-so/.

4. Doreen Dodgen-Magee, *Deviced! Balancing Life and Technology in a Digital World* (Lanham, MD: Rowman & Littlefield, 2018).

5. Elisabeth Kübler-Ross (1969), *On Death and Dying* (New York: Routledge, 2002).

6. M. Katherine Shear, "Grief and Mourning Gone Awry: Pathway and Course of Complicated Grief," *Dialogues in Clinical Neuroscience* 14, no. 2 (June 2012): 119–28. https://www.ncbi.nlm.nih.gov/pmc/articles/PMC3384440/.

7. Amy Stewart, LCSW (October 8, 2020).

8. Cirecie West-Olatunji, "Suicide Prevention in the Age of Coronavirus: Working with Disaster-Affected Clients," *Mental Health Academy 2020 Suicide Prevention Summit* webinar, August 30, 2020.

9. https://ncsacw.samhsa.gov/userfiles/files/SAMHSA_Trauma.pdf.

10. Erika Hayasaki, "How to Remember a Disaster without Being Shattered by It," *Wired*, February 3, 2021, https://www.wired.com/story/remember-disaster-without-being-shattered-ptsd-covid/.

11. Ashley Ceck, "Applying Trauma Informed Best Practices," Gun Sense University Online, https://everytown.docebosaas.com/learn.

12. Julianne Holt-Lunstad, "The Potential Public Health Relevance of Social Isolation and Loneliness: Prevalence, Epidemiology, and Risk Factors," *Public Policy & Aging Report* 27, no. 4 (2017): 127–30, https://doi.org/10.1093/ppar/prx030.

13. "The 'Loneliness Epidemic,'" HRSA (Health Resources & Services Administration), U.S. Department of Health and Human Services, January

10, 2019, https://www.hrsa.gov/enews/past-issues/2019/january-17/loneliness
-epidemic.

14. "The Health Impact of Loneliness: Emerging Evidence and Interventions," NIHCM, October 15, 2021, https://nihcm.org/publications/the
-health-impact-of-loneliness-emerging-evidence-and-interventions.

15. Neil Howe, "Millennials and the Loneliness Epidemic," *Forbes*, May
3, 2019, https://www.forbes.com/sites/neilhowe/2019/05/03/millennials-and
-the-loneliness-epidemic/?sh=61ed58307676.

16. Kasley Killam, "In the Midst of the Pandemic, Loneliness Has Leveled
Out," *Scientific American*, August 18, 2020, https://www.scientificamerican
.com/article/in-the-midst-of-the-pandemic-loneliness-has-leveled-out/.

17. M. Luchetti et al., "The Trajectory of Loneliness in Response to
COVID-19," *American Psychologist* 75, no. 7 (2020): 897–908, https://doi.org
/10.1037/amp0000690.

18. Dunigan Folk et al., "Did Social Connection Decline during the First
Wave of COVID-19? The Role of Extraversion," University of California Press,
July 24, 2020, https://online.ucpress.edu/collabra/article/6/1/37/114469/Did
-Social-Connection-Decline-During-the-First.

19. Killam, "In the Midst of the Pandemic, Loneliness Has Leveled Out."

20. Brian Mastroianni, "More Stressed Than Ever Since COVID-19 Started?
You're Not Alone," Healthline, February 5, 2021, https://www.healthline
.com/health-news/people-feeling-more-stress-now-than-any-point-since-the
-pandemic-began#How-this-stress-is-impacting-our-health.

21. Jon Kabat-Zinn, *Full Catastrophe Living Using the Wisdom of Your Body and
Mind to Face Stress, Pain, and Illness* (New York: Bantam Books, 2013).

22. "2021 Stress in America Post Inauguration Survey Methods," American
Psychological Association, accessed March 7, 2021, https://www.apa.org/news
/press/releases/stress/2021/methodology-january.

23. Karlis, "Is the Pandemic Making Our Social Skills Decay?"

24. "Depression," American Psychological Association, accessed March 7,
2021, https://www.apa.org/topics/depression/.

25. Nirmita Panchal et al., "The Implications of COVID-19 for Mental
Health and Substance Use," KFF, February 10, 2021, https://www.kff.org
/coronavirus-covid-19/issue-brief/the-implications-of-covid-19-for-mental
-health-and-substance-use/.

26. "Anxiety," American Psychological Association, accessed March 7, 2021,
https://www.apa.org/topics/anxiety/.

27. Allen Kim, "Young People's Anxiety Levels Nearly Doubled during
First COVID-19 Lockdown, Study Says," CNN Health, November 25, 2020,

https://www.cnn.com/2020/11/25/health/covid-mental-health-wellness
-trnd/index.html.

28. Ashley Abramson, "Substance Use during the Pandemic," Monitor on Psychology (American Psychological Association), March 1, 2021, https://www.apa.org/monitor/2021/03/substance-use-pandemic.

29. Abramson, "Substance Use during the Pandemic."

30. Jeffrey Kluger, "Domestic Violence Is a Pandemic within the COVID-19 Pandemic," *Time*, February 3, 2021, https://time.com/5928539/domestic -violence-covid-19/.

31. SAMHSA, "Intimate Partner Violence and Child Abuse Considerations during COVID-19."

32. Kristen Rogers, "Mental Health Is One of the Biggest Pandemic Issues We'll Face in 2021," CNN Health, January 4, 2021, https://www.cnn.com /2021/01/04/health/mental-health-during-covid-19-2021-stress-wellness /index.html.

CHAPTER 3

1. "Long-Term Effects of COVID-19," Centers for Disease Control and Prevention, November 13, 2020, https://www.cdc.gov/coronavirus/2019-ncov /long-term-effects.html.

2. David Kessler, "Our Experience of Grief Is Unique as a Fingerprint," Literary Hub, November 15, 2019, https://lithub.com/our-experience-of-grief -is-unique-as-a-fingerprint/.

3. John Ratey, "Welcome to John Ratey M.D. Cambridge, MA," accessed March 8, 2021, http://www.johnratey.com/.

4. Kessler, "Our Experience of Grief Is Unique as a Fingerprint."

CHAPTER 4

1. Jean Spencer, "Checklists for Reopening Business after COVID-19," Workest by Zenefits, May 7, 2020, https://www.zenefits.com/workest/checklists-for -reopening-business-after-covid-19/.

2. David M. DeJoy et al., "Assessing the Impact of Healthy Work Organization Intervention," *Journal of Occupational and Organizational Psychology* 83, no. 1 (2010): 139–65, https://doi.org/10.1348/096317908x398773.

3. Angele Farrell and Patricia Geist-Martin, "Communicating Social Health," *Management Communication Quarterly* 18, no. 4 (2005): 543–92, https://doi.org/10.1177/0893318904273691.

4. Kate Sparks, Brian Faragher, and Cary L. Cooper, "Well-Being and Occupational Health in the 21st Century Workplace," British Psychological Society, *Journal of Occupational and Organizational Psychology* 74, no. 4 (November 2001): 489–509, https://bpspsychub.onlinelibrary.wiley.com/doi/abs/10.1348/096317901167497.

5. Rampalli Prabhakara Raya and Sivapragasam Panneerselvam, "The Healthy Organization Construct: A Review and Research Agenda," *Indian Journal of Occupational and Environmental Medicine* 17, no. 3 (2013): 89–93, https://doi.org/10.4103/0019-5278.130835.

6. Mo Siu-Mei Lee et al., "Relationship between Mental Health and Job Satisfaction among Employees in a Medical Center Department of Laboratory Medicine," *Journal of the Formosan Medical Association* 108, no. 2 (2009): 146–54, https://doi.org/10.1016/s0929-6646(09)60045-0.

CHAPTER 5

1. Jennifer Senior, "Mothers All Over Are Losing It," *New York Times*, February 24, 2021, https://www.nytimes.com/2021/02/24/opinion/covid-pandemic-mothers-parenting.html?referringSource=articleShare.

2. Julia Carmel, "To Hear America's Mothers, We Let Them Scream," *New York Times*, February 6, 2021, https://www.nytimes.com/2021/02/06/insider/primal-scream-section.html.

3. Daniel Siegel, "Making Sense of Your Past," PsychAlive, April 8, 2016, https://www.psychalive.org/the-importance-of-making-sense-of-our-pasts-by-daniel-siegel-m-d/.

4. Leah M. Kuypers, *The Zones of Regulation: A Curriculum Designed to Foster Self-Regulation and Emotional Control* (Santa Clara, CA: Think Social Publishing, 2011).

5. William R. Miller and Stephen Rollnick, *Motivational Interviewing: Helping People Change*, 3rd ed. (New York: Guilford Press, 2013).

6. "Anxiety and Depression in Children," Centers for Disease Control and Prevention, December 2, 2020, https://www.cdc.gov/childrensmentalhealth/depression.html.

7. "Anxiety and Depression in Children," Centers for Disease Control and Prevention.

8. "Resilience," Center on the Developing Child at Harvard University, August 17, 2020, https://developingchild.harvard.edu/science/key-concepts/resilience/.

9. Katie Hurley, "Build Resilience in Children: Strategies to Strengthen Your Kids," Psycom—A Mental Health Resource since 1996, November 24, 2020, https://www.psycom.net/build-resilience-children.

CHAPTER 6

1. "Core Values List," James Clear website, June 12, 2018, https://jamesclear.com/core-values.

CHAPTER 7

1. James Clear, "This Zen Concept Will Help You Stop Being a Slave to Old Beliefs," James Clear website, accessed January 11, 2021, https://jamesclear.com/shoshin.

2. Shunryu Suzuki, *Zen Mind, Beginner's Mind: Informal Talks on Zen Meditation and Practice* (Boulder, CO: Weatherhill, 1970).

3. Leo Babauta, "Approaching Life with Beginner's Mind," Zen Habits website, accessed January 11, 2021, https://zenhabits.net/beginner/.

CHAPTER 8

1. https://guilfordjournals.com/doi/1.1521/jscp.2018.37.10.751.

2. Edita Poljac et al., "New Perspectives on Human Multitasking," *Psychological Research* 82, nos. 1–3 (January 18, 2018), doi:10.1007/s00426-018-0970-2.

3. David M. Sanbonmatsu et al., "Who Multi-Tasks and Why? Multi-Tasking Ability, Perceived Multi-Tasking Ability, Impulsivity, and Sensation Seeking," *PLoS ONE* 8, no. 1 (January 23, 2013), doi:10.1371/journal.pone.0054402.

4. Kevin P. Madore and Anthony D. Wagner, "Multicosts of Multitasking," *Cerebrum*, April 1, 2019, https://www.ncbi.nlm.nih.gov/pmc/articles/PMC7075496/#__ffn_sectitle.

5. Melina R. Uncapher et al., "Media Multitasking and Cognitive, Psychological, Neural, and Learning Differences," *Pediatrics* 140, no. 2 (November 2017): S62–S66, doi:10.1542/peds.2016-1758D.

6. Melina R. Uncapher, Monica K. Thieu, and Anthony D. Wagner, "Media Multitasking and Memory: Differences in Working Memory and Long-Term Memory," *Psychonomic Bulletin & Review* 23, no. 2 (April 2016): 483–90, doi:10.3758/s13423-015-0907-3.

7. Kep Kee Loh and Ryota Kanai, "Higher Media Multi-Tasking Activity Is Associated with Smaller Gray-Matter Density in the Anterior Cingulate Cortex," *PLoS ONE* 9, no. 9 (September 24, 2014): e106698, doi:10.1371/journal.pone.0106698; and Eyal Ophir, Clifford Nass, and Anthony D. Wagner, "Cognitive Control in Media Multitaskers," *Proceedings of the National Academy of Sciences of the USA* 106, no. 37 (September 15, 2009): 15583–87, doi:10.1073/pnas.0903620106.

8. Brandon C. W. Ralph et al., "Media Multitasking and Behavioral Measures of Sustained Attention," *Attention, Perception, & Psychophysics* 77, no. 2 (February 2015): 390–401, doi:10.3758/s13414-014-0771-7.

9. Matthew S. Cain et al., "Media Multitasking in Adolescence," *Psychonomic Bulletin & Review* 23, no. 6 (December 2016): 1932–41, doi:10.3758/s13423-016-1036-3.

10. Mark W. Becker, Reem Alzahabi, and Christopher J. Hopwood, "Media Multitasking Is Associated with Symptoms of Depression and Social Anxiety," *Cyberpsychology, Behavior, and Social Networking* 16, no. 2 (February 2013): 132–35, doi:10.1089/cyber.2012.0291.

11. Roy Pea et al., "Media Use, Face-to-Face Communication, Media Multitasking, and Social Well-Being among 8- to 12-Year-Old Girls," *Developmental Psychology* 48, no. 2 (March 2012): 327–36, doi:10.1037/a0027030.

12. Shalena Srna, Rom Y. Schrift, and Gal Zauberman, "The Illusion of Multitasking and Its Positive Effect on Performance," *Psychological Science* 29, no. 12 (October 24, 2018): 1942–55, doi:10.1177/0956797618801013.

13. "The 'Loneliness Epidemic,'" Health Resources & Services Administration, U.S. Department of Health and Human Services, accessed January 16, 2021, https://www.hrsa.gov/enews/past-issues/2019/january-17/loneliness-epidemic.

14. Yuval Palgi et al., "The Loneliness Pandemic: Loneliness and Other Concomitants of Depression, Anxiety, and Their Comorbidity during the COVID-19 Outbreak," *Journal of Affective Disorders* 275 (October 1, 2020): 109–11, doi:10.1016/j.jad.2020.06.036.

15. Philip Jefferies and Michael Ungar, "Social Anxiety in Young People: A Prevalence Study in Seven Countries," *PLoS ONE* 15, no. 9 (September 17, 2020): e0239133, doi:10.1371/journal.pone.0239133.

16. L. Sareen and M. Stein, "A Review of the Epidemiology and Approaches to the Treatment of Social Anxiety Disorder," *Drugs* 59, no. 3 (March 2000): 497–509, doi: 10.2165/00003495-200059030-00007.

17. Ahmet Koyuncu et al., "Comorbidity in Social Anxiety Disorder: Diagnostic and Therapeutic Challenges," *Drugs Context* 8 (April 2, 2019): 212573, doi:10.7573/dic.212573.

18. Judith Dams et al., "Excess Costs of Social Anxiety Disorder in Germany," *Journal of Affective Disorders* 213 (April 15, 2017): 23-29, doi:10.1016/j.jad.2017.01.041.

19. Daniel Siegel, *Mindsight: The New Science of Personal Transformation* (New York: Bantam Books, 2010).

CHAPTER 9

1. Marshall B. Rosenberg and Lucy Leu, "NVC Instruction Self-Guide," Center for Nonviolent Communication, accessed March 8, 2021, https://www.cnvc.org/online-learning/nvc-instruction-guide/nvc-instruction-guide.

2. Audre Lorde, *A Burst of Light: and Other Essays* (Mineola, NY: Ixia Press, 2017).

BIBLIOGRAPHY

"2021 Stress in America Post Inauguration Survey Methods." American Psychological Association. Accessed March 7, 2021. https://www.apa.org/news/press/releases/stress/2021/methodology-january.

Abramson, Ashley. "Substance Use during the Pandemic." Monitor on Psychology, American Psychological Association, March 1, 2021. https://www.apa.org/monitor/2021/03/substance-use-pandemic.

Adichie, Chimamanda Ngozi. "The Danger of a Single Story." TED Ideas Worth Spreading, July 2009. Video, 18:33. https://www.ted.com/talks/chimamanda_ngozi_adichie_the_danger_of_a_single_story.

"Anxiety and Depression in Children." Centers for Disease Control and Prevention, December 2, 2020. https://www.cdc.gov/childrensmentalhealth/depression.html.

"Anxiety." American Psychological Association. Accessed March 7, 2021. https://www.apa.org/topics/anxiety/.

Babauta, Leo. "Approaching Life with Beginner's Mind." Zen Habits website. Accessed January 11, 2021. https://zenhabits.net/beginner/.

Becker, Mark W., Reem Alzahabi, and Christopher J. Hopwood. "Media Multitasking Is Associated with Symptoms of Depression and Social Anxiety." Cyberpsychology, Behavior, and Social Networking 16, no. 2 (February 2013): 132–35. doi:10.1089/cyber.2012.0291.

Berger, Michele W. "Social Media Use Increases Depression and Loneliness." Penn Today, University of Pennsylvania, November 9, 2018. https://penntoday.upenn.edu/news/social-media-use-increases-depression-and-loneliness.

Cain, Matthew S., Julia A. Leonard, John D. E. Gabrieli, and Amy S. Finn. "Media Multitasking in Adolescence." Psychonomic Bulletin & Review 23, no. 6 (December 2016): 1932–41. doi:10.3758/s13423-016-1036-3.

Carmel, Julia. "To Hear America's Mothers, We Let Them Scream." New York Times, February 6, 2021. https://www.nytimes.com/2021/02/06/insider/primal-scream-section.html.

Ceck, Ashley. "Applying Trauma Informed Best Practices." Gun Sense University Online, August 2020. https://everytown.docebosaas.com/learn.

Clear, James. "This Zen Concept Will Help You Stop Being a Slave to Old Beliefs." James Clear website. Accessed January 11, 2021. https://jamesclear.com/shoshin.

"Concept of Trauma and Guidance for a Trauma-Informed Approach." SAMHSA's Trauma and Justice Strategic Initiative, July 2014. https://ncsacw.samhsa.gov/userfiles/files/SAMHSA_Trauma.pdf.

"Controlling Anger before It Controls You." American Psychological Association. Accessed March 8, 2021. https://www.apa.org/topics/anger/control.

Dams, Judith, Hans-Helmut König, Florian Bleibler, Jürgen Hoyer, Jörg Wiltink, Manfred E. Beutel et al. "Excess Costs of Social Anxiety Disorder in Germany." *Journal of Affective Disorders* 213 (April 15, 2017): 23–29. doi:10.1016/j.jad.2017.01.041.

DeJoy, David M., Mark G. Wilson, Robert J. Vandenberg, Allison L. McGrath-Higgins, and Shannon C. Griffin-Blake. "Assessing the Impact of Healthy Work Organization Intervention." *Journal of Occupational and Organizational Psychology* 83, no. 1 (2010): 139–65. https://doi.org/10.1348/096317908x398773.

"Depression." American Psychological Association. Accessed March 7, 2021. https://www.apa.org/topics/depression/.

Dodgen-Magee, Doreen. *Deviced! Balancing Life and Technology in a Digital World.* Lanham, MD: Rowman & Littlefield, 2018.

Dodgen-Magee, Doreen. "Grumpy, Sad, and Anxious: Week 9 of Quarantine." *Psychology Today* (blog), May 13, 2020. https://www.psychologytoday.com/us/blog/deviced/202005/grumpy-sad-and-anxious-week-9-quarantine.

Farrell, Angele, and Patricia Geist-Martin. "Communicating Social Health." *Management Communication Quarterly* 18, no. 4 (2005): 543–92. https://doi.org/10.1177/0893318904273691.

Folk, Dunigan, Karynna Okabe-Miyamoto, Elizabeth Dunn, Sonja Lyubomirsky, and Brent Donnellan. "Did Social Connection Decline during the First Wave of COVID-19? The Role of Extraversion." University of California Press, July 24, 2020. https://online.ucpress.edu/collabra/article/6/1/37/114469/Did-Social-Connection-Decline-During-the-First.

Hayasaki, Erika. "How to Remember a Disaster without Being Shattered by It." *Wired*, February 3, 2021. https://www.wired.com/story/remember-disaster-without-being-shattered-ptsd-covid/.

"Health Equity Considerations and Racial and Ethnic Minority Groups." Centers for Disease Control and Prevention, updated April 19, 2021.

https://www.cdc.gov/coronavirus/2019-ncov/community/health-equity /race-ethnicity.html.

"The Health Impact of Loneliness: Emerging Evidence and Interventions" Webinar, NIHCM (National Institute for Healthcare Management), October 15, 2021. https://nihcm.org/publications/the-health-impact-of-loneliness -emerging-evidence-and-interventions.

Holt-Lunstad, Julianne. "The Potential Public Health Relevance of Social Isolation and Loneliness: Prevalence, Epidemiology, and Risk Factors." *Public Policy & Aging Report* 27, no. 4 (2017): 127–30. https://doi.org/10.1093 /ppar/prx030.

Howe, Neil. "Millennials and the Loneliness Epidemic." *Forbes*, May 3, 2019. https://www.forbes.com/sites/neilhowe/2019/05/03/millennials-and-the -loneliness-epidemic/?sh=61ed58307676.

Hurley, Katie. "Build Resilience in Children: Strategies to Strengthen Your Kids." Psycom—A Mental Health Resource since 1996, November 24, 2020. https://www.psycom.net/build-resilience-children.

"Intimate Partner Violence and Child Abuse Considerations during COVID-19." SAMHSA, March 7, 2020. https://www.samhsa.gov/sites /default/files/social-distancing-domestic-violence.pdf.

Jefferies, Philip, and Michael Ungar. "Social Anxiety in Young People: A Prevalence Study in Seven Countries." *PLoS ONE* 15, no. 9 (September 17, 2020): e0239133. doi:10.1371/journal.pone.0239133.

Jha, Rega. "How Privileged Are You?" Buzzfeed, April 10, 2014. https:// www.buzzfeed.com/regajha/how-privileged-are-you.

Kabat-Zinn, Jon. *Full Catastrophe Living Using the Wisdom of Your Body and Mind to Face Stress, Pain, and Illness.* New York: Bantam Books, 2013.

Karlis, Nicole. "Is the Pandemic Making Our Social Skills Decay? Psychologists Think So." *Salon*, February 5, 2021. https://www.salon.com/2021/02/04 /is-the-pandemic-making-our-social-skills-decay-psychologists-think-so/.

Kassinove, Howard, ed. *Anger Disorders: Definition, Diagnosis and Treatment.* Washington, DC: Taylor & Francis, 1998.

Kessler, David. "Our Experience of Grief Is Unique as a Fingerprint." Literary Hub (blog), November 15, 2019. https://lithub.com/our-experience-of -grief-is-unique-as-a-fingerprint/.

Killam, Kasley. "In the Midst of the Pandemic, Loneliness Has Leveled Out." *Scientific American*, August 18, 2020. https://www.scientificamerican.com /article/in-the-midst-of-the-pandemic-loneliness-has-leveled-out/.

Kim, Allen. "Young People's Anxiety Levels Nearly Doubled during First COVID-19 Lockdown, Study Says." CNN Health, November 25, 2020.

https://www.cnn.com/2020/11/25/health/covid-mental-health-wellness
-trnd/index.html.

Kluger, Jeffrey. "Domestic Violence Is a Pandemic within the COVID-19 Pandemic." *Time*, February 3, 2021. https://time.com/5928539/domestic -violence-covid-19/.

Koyuncu, Ahmet, Ezgi İnce, Erhan Ertekin, and Raşit Tükel. "Comorbidity in Social Anxiety Disorder: Diagnostic and Therapeutic Challenges." *Drugs Context* 8 (April 2, 2019): 212573. doi:10.7573/dic.212573.

Kübler-Ross, Elisabeth. *On Death and Dying: Questions and Answers on Death and Dying, On Life after Death.* New York: Routledge, 2002.

Kuypers, Leah M. *The Zones of Regulation: A Curriculum Designed to Foster Self-Regulation and Emotional Control.* Santa Clara, CA: Think Social Publishing, 2011.

Lee, Mo Siu-Mei, Ming-Been Lee, Shih-Cheng Liao, and Fu-Tien Chiang. "Relationship between Mental Health and Job Satisfaction among Employees in a Medical Center Department of Laboratory Medicine." *Journal of the Formosan Medical Association* 108, no. 2 (2009): 146–54. https://doi.org/10.1016 /s0929-6646(09)60045-0.

Loh, Kep Kee, and Ryota Kanai. "Higher Media Multi-Tasking Activity Is Associated with Smaller Gray-Matter Density in the Anterior Cingulate Cortex." *PLoS ONE* 9, no. 9 (September 24, 2014): e106698. doi:10.1371 /journal.pone.0106698.

"The 'Loneliness Epidemic.'" HRSA (Health Resources & Services Administration), U.S. Department of Health and Human Services, January 10, 2019. https://www.hrsa.gov/enews/past-issues/2019/january-17/loneliness -epidemic.

"Long-Term Effects of COVID-19." Centers for Disease Control and Prevention, November 13, 2020. https://www.cdc.gov/coronavirus/2019-ncov/ long-term-effects.html.

Lorde, Audre. *A Burst of Light: and Other Essays.* Mineola, NY: Ixia Press, 2017.

Luchetti, M., J. H. Lee, D. Aschwanden, A. Sesker, J. E. Strickhouser, A. Terracciano, and A. R. Sutin. "The Trajectory of Loneliness in Response to COVID-19." *American Psychologist* 75, no. 7 (2020): 897–908. http:// dx.doi.org/10.1037/amp0000690.

Madore, Kevin P., and Anthony D. Wagner. "Multicosts of Multitasking," National Center for Biotechnology Information, *Cerebrum*, April 1, 2019. https://www.ncbi.nlm.nih.gov/pmc/articles/PMC7075496/#__ffn_sectitle.

Mastroianni, Brian. "More Stressed Than Ever since COVID-19 Started? You're Not Alone." Healthline, February 5, 2021. https://www.healthline

.com/health-news/people-feeling-more-stress-now-than-any-point-since -the-pandemic-began#How-this-stress-is-impacting-our-health.

Miller, William R., and Stephen Rollnick. *Motivational Interviewing: Helping People Change*, 3rd ed. New York: Guilford Press, 2013.

Ophir, Eyal, Clifford Nass, and Anthony D. Wagner. "Cognitive Control in Media Multitaskers." *Proceedings of the National Academy of Sciences of the USA* 106, no. 37 (September 15, 2009): 15583–87. doi:10.1073/pnas.0903620106.

Palgi, Yuval, Amit Shrira, Lia Ring, Ehud Bodner, Sharon Avidor, Yoav Bergman, Sara Cohen-Fridel, Shoshi Keisari, and Yaakov Hoffman. "The Loneliness Pandemic: Loneliness and Other Concomitants of Depression, Anxiety, and Their Comorbidity during the COVID-19 Outbreak." *Journal of Affective Disorders* 275 (October 1, 2020): 109–11. doi:10.1016/j.jad.2020.06.036.

Panchal, Nirmita, Rabah Kamal, Cynthia Cox, and Rachel Garfield. "The Implications of COVID-19 for Mental Health and Substance Use." KFF (Kaiser Family Foundation), February 10, 2021. https://www.kff.org/coronavirus -covid-19/issue-brief/the-implications-of-covid-19-for-mental-health-and -substance-use/.

Pea, Roy, Clifford Nass, Lyn Meheula, Marcus Rance, Aman Kumar, Holden Bamford, Matthew Nass, Aneesh Simha, Benjamin Stillerman, Steven Yang, and Michael Zhou. "Media Use, Face-to-Face Communication, Media Multitasking, and Social Well-Being among 8- to 12-Year-Old Girls." *Developmental Psychology* 48, no. 2 (March 2012): 327–36. doi:10.1037/a0027030.

Poljac, Edita, Andrea Kiesel, Iring Koch, and Hermann Müller. "New Perspectives on Human Multitasking." *Psychological Research* 82, nos. 1–3 (January 18, 2018). doi:10.1007/s00426-018-0970-2.

Ralph, Brandon C. W., David R. Thomson, Paul Seli, Jonathan S. A. Carriere, and Daniel Smilek. "Media Multitasking and Behavioral Measures of Sustained Attention." *Attention, Perception, & Psychophysics* 77, no. 2 (February 2015): 390–401. doi:10.3758/s13414-014-0771-7.

Ratey, John. "Welcome to John Ratey M.D. Cambridge, MA." Accessed March 8, 2021. http://www.johnratey.com/.

Raya, Rampalli Prabhakara, and Sivapragasam Panneerselvam. "The Healthy Organization Construct: A Review and Research Agenda." *Indian Journal of Occupational and Environmental Medicine* 17, no. 3 (2013): 89–93. https://doi .org/10.4103/0019-5278.130835.

"Resilience." Center on the Developing Child at Harvard University, August 17, 2020. https://developingchild.harvard.edu/science/key-concepts/resilience/.

Rogers, Kristen. "Mental Health Is One of the Biggest Pandemic Issues We'll Face in 2021." CNN Health, January 4, 2021. https://www.cnn.com/2021

/01/04/health/mental-health-during-covid-19-2021-stress-wellness/index
.html.

Sanbonmatsu, David M., David L. Strayer, Nathan Medeiros-Ward, and Jason
M. Watson. "Who Multi-Tasks and Why? Multi-Tasking Ability, Perceived
Multi-Tasking Ability, Impulsivity, and Sensation Seeking." *PLoS ONE* 8,
no. 1 (January 23, 2013). doi:10.1371/journal.pone.0054402.

Sareen, L., and M. Stein. "A Review of the Epidemiology and Approaches to
the Treatment of Social Anxiety Disorder." *Drugs* 59, no. 3 (March 2000):
497–509. doi: 10.2165/00003495-200059030-00007.

Senior, Jennifer. "Mothers All Over Are Losing It." *New York Times*, February
24, 2021. https://www.nytimes.com/2021/02/24/opinion/covid-pandemic
-mothers-parenting.html?referringSource=articleShare.

Shear, M. Katherine. "Grief and Mourning Gone Awry: Pathway and Course
of Complicated Grief." *Dialogues in Clinical Neuroscience* 14, no. 2 (June 2012):
119–28. https://www.ncbi.nlm.nih.gov/pmc/articles/PMC3384440/.

Siegel, Daniel. "Making Sense of Your Past." PsychAlive, April 8, 2016. https://
www.psychalive.org/the-importance-of-making-sense-of-our-pasts-by
-daniel-siegel-m-d/.

Siegel, Daniel. *Mindsight: The New Science of Personal Transformation*. New York:
Bantam Books, 2010.

"Social deprivation." APA Dictionary of Psychology. American Psychological Association. Accessed March 8, 2021. https://dictionary.apa.org/social
-deprivation.

Sparks, Kate, Brian Faragher, and Cary L. Cooper. "Well-Being and Occupational Health in the 21st Century Workplace." British Psychological Society,
Journal of Occupational and Organizational Psychology 74, no. 4 (November 2001):
489–509. https://bpspsychub.onlinelibrary.wiley.com/doi/abs/10.1348
/096317901167497.

Spencer, Jean. "Checklists for Reopening Business after COVID-19." Workest
by Zenefits, May 7, 2020. https://www.zenefits.com/workest/checklists-for
-reopening-business-after-covid-19/.

Srna, Shalena, Rom Y. Schrift, and Gal Zauberman. "The Illusion of Multitasking and Its Positive Effect on Performance." *Psychological Science* 29, no. 12
(October 24, 2018): 1942–55. doi:10.1177/0956797618801013.

Stewart, Amy. "COVID Survivors for Change." *COVID Connections Lecture.*
Lecture, October 8, 2020.

Suzuki, Shunryu. *Zen Mind, Beginner's Mind: Informal Talks on Zen Meditation
and Practice*. Boulder, CO: Weatherhill, 1970.

True, Sarah, Nancy Ochieng, Juliette Cubanski, Priya Chidambaram, and
Tricia Neuman. "Overlooked and Undercounted: The Growing Impact

of COVID-19 on Assisted Living Facilities." KFF (Kaiser Family Foundation), September 1, 2020. https://www.kff.org/coronavirus-covid-19/issue-brief/overlooked-and-undercounted-the-growing-impact-of-covid-19-on-assisted-living-facilities/.

Uncapher, Melina R., Lin Lin, Larry D. Rosen, Heather L. Kirkorian, Naomi S. Baron, Kira Bailey, Joanne Cantor, David L. Strayer, Thomas D. Parsons, and Anthony D. Wagner. "Media Multitasking and Cognitive, Psychological, Neural, and Learning Differences." *Pediatrics* 140, supplement no. 2 (November 2017): S62–S66. doi:10.1542/peds.2016-1758D.

Uncapher, Melina R., Monica K. Thieu, and Anthony D. Wagner. "Media Multitasking and Memory: Differences in Working Memory and Long-Term Memory." *Psychonomic Bulletin & Review* 23, no. 2 (April 2016): 483–90. doi:10.3758/s13423-015-0907-3.

West-Olatunji, Cirecie. "Suicide Prevention in the Age of Coronavirus: Working with Disaster-Affected Clients." *Mental Health Academy 2020 Suicide Prevention Summit*, American Mental Health Counselors, webinar, August 30, 2020.

"What Is Privilege." YMCA of Greater Cleveland, Ohio, March 4, 2019. https://www.ywcaofcleveland.org/blog/2019/03/04/what-is-privilege/.

INDEX

Page references for textboxes are italicized.

ABOUT THE AUTHOR

Doreen Dodgen-Magee is an author, psychologist, and speaker. Her first book, *Deviced! Balancing Life and Technology in a Digital World* (Rowman & Littlefield), was awarded the 2018 Gold Medal for Psychology by the Nautilus Book Awards. Doreen's writing has appeared in the *Washington Post, Chicago Tribune, Utne Reader, Psychology Today,* and *Health* magazine as well as other popular press outlets. She has been interviewed and quoted in articles for the *New York Times, Time* magazine, and *The Guardian*. Doreen is a Senior Survivor Fellow with Everytown for Gun Safety and a group facilitator with COVID Survivors for Change. She lives in Portland, Oregon, with her husband and near her grown children and can be found at www.doreendm.com.

CPSIA information can be obtained
at www.ICGtesting.com
Printed in the USA
BVHW040422060122
624073BV00004B/9

9 781538 160275